THE WHALERS

GREENLAND

BAFFIN
BAY

ICELAND

ARCTIC CIRCLE

ASIA

EUROPE

Liverpool
Milford Haven
London

40° N.

NOVA
SCOTIA

Boston
New Bedford
Nantucket
Martha's Vineyard
Mystic
Long Island

GULF STREAM

AZORES

Charleston
Savannah

BERMUDA ISLANDS

WEST INDIES

CAPE VERDE
ISLANDS

GUINEA COAST

AFRICA

EQUATOR

SOUTH
AMERICA

Callao

BRAZIL
GROUNDS

ATLANTIC OCEAN

INDIAN OCEAN

Rio de Janeiro

Valparaiso

Talcahuano

CAPE OF GOOD HOPE

40° S.

FALKLAND
ISLANDS

CAPE HORN

ANTARCTIC CIRCLE

LAHAINA

The Seafarers THE WHALERS

The Cover: Horror-struck whalemen leap for their lives as an enraged sperm whale crushes their boat in its jaws in this watercolor painted in the 1840s by an unknown artist. A dangerous foe, the sperm whale had "murderous jaws full of great cruel teeth," wrote one whaleman, "and a throat that could take in a man."

The Title Page: Scavenging sea gulls anticipate a bloody kill in this classic scene of whaling carved on the tooth of a sperm whale in the 1830s. Scrimshaw, the art of carving whale teeth and bone, was a favorite pastime of whalemen (*pages 126-131*), and this superb example shows the fine hand of a master scrimshander.

The Seafarers

THE WHALERS

by A. B. C. Whipple

AND THE EDITORS OF TIME-LIFE BOOKS

TIME-LIFE BOOKS, ALEXANDRIA, VIRGINIA

TIME-LIFE BOOKS INC.

MANAGING EDITOR: Jerry Korn
Executive Editor: David Maness
Assistant Managing Editors: Dale M. Brown (planning).
George Constable. Martin Mann. John Paul Porter
Art Director: Tom Suzuki
Chief of Research: David L. Harrison
Director of Photography: Robert G. Mason
Assistant Art Director: Arnold C. Holeywell
Assistant Chief of Research: Carolyn L. Sackett
Assistant Director of Photography: Dolores A. Littles

CHAIRMAN: Joan D. Manley
President: John D. McSweeney
Executive Vice Presidents: Carl G. Jaeger.
John Steven Maxwell. David J. Walsh
Vice Presidents: George Artandi (comptroller);
Stephen L. Bair (legal counsel); Peter G. Barnes;
Nicholas Benton (public relations); John L. Canova;
Beatrice T. Dobie (personnel); Carol Flaumenhaft
(consumer affairs); Nicholas J. C. Ingleton (Asia);
James L. Mercer (Europe/South Pacific); Herbert Sorkin
(production); Paul R. Stewart (marketing)

The Seafarers

Editorial Staff for The Whalers:
Editor: George G. Daniels
Picture Editor: John Conrad Weiser
Designer: Herbert H. Quarmby
Text Editors: Anne Horan, Sterling Seagrave
Staff Writers: William C. Banks. Carol Dana.
Stuart Gannes, Gus Hedberg
Chief Researcher: Charlotte A. Quinn
Researchers: Patti H. Cass, Philip Brandt George.
W. Mark Hamilton, Elizabeth L. Parker.
Trudy W. Pearson, Kathleen Shortall
Art Assistant: Michelle René Clay
Editorial Assistant: Adrienne George

Editorial Production
Production Editor: Douglas B. Graham
Operations Manager: Gennaro C. Esposito.
Gordon E. Buck (assistant)
Assistant Production Editor: Feliciano Madrid
Quality Control: Robert L. Young (director). James J. Cox
(assistant). Daniel J. McSweeney. Michael G. Wight
(associates)
Art Coordinator: Anne B. Landry
Copy Staff: Susan B. Galloway (chief). Sheirazada Hann.
Elise D. Ritter. Celia Beattie
Picture Department: Marguerite Johnson.
Nancy Cromwell Scott

Correspondents: Elisabeth Kraemer (Bonn); Margot
Hapgood. Dorothy Bacon (London); William Lyon
(Madrid); Susan Jonas. Lucy T. Voulgaris (New York);
Maria Vincenza Aloisi. Josephine du Brusle (Paris); Ann
Natanson (Rome).
Valuable assistance was provided by Jay Brennan. Asia
Editor. Tokyo. The editors also wish to thank: Jenny
Hovinga (Amsterdam); Enid Farmer. Sue Wymelenberg
(Boston); Karen Horton (Honolulu); Karin B. Pearce
(London); Carolyn T. Chubet. Miriam Hsia. Christina
Lieberman (New York); Dag Christensen (Oslo); Mimi
Murphy (Rome); Janet Zich (San Francisco); Peter Allen
(Sydney); Katsuko Yamazaki (Tokyo).

The editors are indebted to Champ Clark. Barbara Hicks
and Katie Hooper McGregor for their help in the
preparation of this book.

The Author:
A lifelong student of the sea and maritime affairs, A.B.C. Whipple is descended from a family of New England seafarers and is himself an inveterate sailor. He has written nine books on ships and the sea, including *Yankee Whalers in the South Seas*. He is a former Assistant Managing Editor of Time-Life Books, and the author of *Fighting Sail* in the Seafarers series.

The Consultants:
John Horace Parry, Gardiner Professor of Oceanic History and Affairs at Harvard University, is a renowned maritime historian and the author of many distinguished books, including *The Spanish Seaborne Empire* and *The Discovery of the Sea*.

Richard C. Kugler, a descendant of New England whaling masters, is director of the New Bedford Whaling Museum in Massachusetts. An eminent authority on the subject, he has written "New Bedford and Old Dartmouth: A Portrait of a Region's Past."

Edouard A. Stackpole is the director of the Peter Foulger Museum on Nantucket Island and was formerly the curator of the Marine Historical Association at Mystic Seaport, Connecticut.

William Avery Baker is curator of the Hart Nautical Museum at the Massachusetts Institute of Technology. He has drawn the reconstruction plans for many historically important vessels, including *Mayflower II*, the frigate *Essex* and a 25-foot whaleboat.

Thomas J. McIntyre is a staff assistant in the Office of Marine Mammals and Endangered Species at the National Oceanographic and Atmospheric Administration in Washington, D.C. Cetology—the study of whales—is his specialty, and in 1976 he completed a research tour aboard a Japanese whaleship in the Antarctic.

For information about any Time-Life book. please write:
Reader Information. Time-Life Books.
541 North Fairbanks Court. Chicago. Illinois 60611.

TIME-LIFE is a trademark of Time Incorporated U.S.A.

Library of Congress Cataloguing in Publication Data
Whipple. Addison Beecher Colvin. 1918-
   The whalers.
   (The seafarers; v.8)
   Bibliography: p.
   Includes index.
   1. Whaling—History.   I. Time-Life
Books.   II. Title.   III. Series.
SH381.W54   338.3'72'9509034   78-31228
ISBN 0-8094-2672-2
ISBN 0-8094-2671-4 lib. bdg.

Contents

Yankee scourges of "the great Leviathan"

It was an obsession of mankind in the 19th Century to pit itself against nature, and no greater natural adversary could be found than the mighty whale. What now seems cruel seemed valiant to the Yankees of old New England: mortal men in frail boats waging desperate battle against a beast they called "the great Leviathan."

On the side of the whalemen was the virtue of economic benefit—whale oil lighted the lamps of the world and lubricated the gears of the dawning Industrial Age. But there was one even stronger motive: to slay the whale was to act in God's name. In a time when fundamentalist beliefs were widespread, the huge mammal was perceived as the incarnation of evil, "the gliding great demon of the seas of life," as Herman Melville's Captain Ahab described Moby Dick. New England preachers were fond of quoting with certitude from the Book of Isaiah, which stated that "the Lord with his sore, and great, and strong sword, shall punish Leviathan the piercing serpent, even Leviathan that crooked serpent; and he shall slay the dragon that is in the sea."

Moral and economic rationales aside, Americans thrilled to the fantastic hunt itself, waged by men of daring under awesome conditions in the vast solitude of the sea. In the supreme age of American whaling, from 1820 to the Civil War, Yankees in their whalers ventured to the very ends of the Atlantic and the Pacific, charting the seas of the world as no explorers had before them. They were often gone for years at a stretch; then they would sail home laden with a treasure of oil, bone, ambergris and spermaceti.

Owners and captains became rich and joined the growing American mercantile aristocracy. On the other hand, seamen generally earned only a scant living. Yet life ashore was unbearably dull compared with the heroic pursuit of the Leviathan. Whaling men knew that many of their company would perish in ships smashed by tropical typhoons or crushed by arctic ice. And there were many terrible ways in which a maddened whale could destroy an entire boatload of men. But those who returned to tell the legends of the incredible contest became themselves a legend.

A fleet of American whalers spans the Bering Strait, portraying the various stages of arctic whaling in this 1871 lithograph. Just below Siberia's East Cape at far left a ship lies crushed in ice. Nearby, whaleboats tow a dead whale to a ship. Distant whalers emit plumes of smoke from tryworks. At center a ship winches aboard blubber strips as a laden whaler passes homeward bound. On the Alaskan side at right, several whaleboats pursue their quarry, while the crew of a whaler hauls on board a whale's head.

Preparing to lance a harpooned whale behind its flipper, an officer braces himself in the whaleboat's bow while the original harpooneer takes over the steering oar. Gulls peck at marine life clinging to the wounded animal. In the background the whaler's crew is busy "cutting in," or flensing, another whale, floating dead on the surface. At far right in this 19th Century French aquatint, another whaleboat hauls in a third victim, which bears a flag of capture drooping from a pole thrust into its back.

While two other whaleboats maneuver cautiously off to the sides, a whaleman brings his boat in close to a thrashing sperm whale and delivers the coup de grâce with an explosive device of the mid-1800s known as a shoulder gun. This 1852 American etching is highly romanticized: whalemen normally would wait until the creature calmed down before venturing this close, and though whaleboats were small and vulnerable, they were not the flimsy cockleshells shown here awash in the seas.

Revenging itself upon its tormentors, an infuriated sperm whale smashes their whaleboat, toppling some of the crewmen while others cling precariously to the wreckage or lie trapped in the tangle of whale line. From the circling sea birds, whalemen could tell roughly where a whale would surface after it had sounded in search of giant squid but, as shown so dramatically in this 1835 French aquatint, it was almost impossible to escape if the angry victim chose to surface right under the boat.

Hauling a captured whale up to the side of the ship, a cluster of crewmen in the bow bend to the windlass as others in a whaleboat help guide the behemoth's carcass. Already sailors amidships are lowering the cutting platform, which will enable them to stand outboard of the dead whale and slice strips of blubber free as they fend off marauding sharks. But there are a couple of errors in this 1865 watercolor: the whale is shown improperly hitched and the whaleboats in the davits are of the wrong shape.

Chapter 1

"Into the charmed, churned circle"

elson Haley never forgot the January afternoon in 1851 when he stood high in the crosstrees of the *Charles W. Morgan* searching the horizon for sperm whales. He was only 18 years old, but already he was an experienced whaleman. He had made one previous voyage, during which he had proved himself so keen of eye and strong of arm that he had risen to harpooner, as he was then called, second only to the captain and mates in the whale crew hierarchy. Thus far on this voyage "Nelt" Haley had made the first strike on one whale. But it was of average size, and he was anxious to prove his right to his rank once again.

Suddenly Haley stiffened and cried: "Great God, there lays a sperm whale as big as a mountain!" Down below, every eye was instantly turned in the direction Haley was pointing. But there was nothing to be seen. Either young Haley's eyes were playing tricks on him, or the great whale had dived before anyone else could spot it. The captain had to make a decision. With scarcely a moment's hesitation he backed Haley, and ordered four boats launched in pursuit.

The *Morgan* was a few hundred miles north of New Zealand, in a flat calm, broken only by the long Pacific swells. "White-ash weather," whalemen disgustedly called it. With no breeze, the whaleboats' canvas sails were useless, and the men had to bend to the supple white-ash oars.

To Haley's vast relief, as they rowed forward, the huge whale broke the surface again, for all to see, in exactly the spot he had indicated. But then it dived before any of the boats could reach it. Where had it gone? Where would it reappear next?

The captain, with two other boats following his lead, turned and raced back toward the *Morgan*, gambling that the whale would rise in that direction. But Haley thought they were overshooting the mark, and he said so to the second mate, who was in command of his boat. The mate agreed with him—but reluctantly, because he knew that he was expected to follow the captain. The lone whaleboat sat in the water, its five oars dripping, as the other boats rowed farther and farther away. Haley was betting that the whale had gone straight down and would resurface close by. Knowing how angry the captain could get, everyone in the second mate's boat prayed that Haley was right.

Haley never forgot the scene as he and his companions sat through the long minutes of waiting. The whaleboat floated like a chip on the immensity of the ocean. The other boats were soon so far off that Haley could see them only when his boat rose with the rolling sea. Now and then an oarlock would creak or a man would cough. Haley was in the bow, ready to seize his harpoon; the second mate was at the stern, leaning on his long steering oar, and the four oarsmen between them shaded their eyes from the dazzling late-afternoon sun and squinted across the still ocean.

A flock of petrels skimmed the water in wide circles, waiting for the whale to surface so they could feast on bits of whatever the enormous beast had captured. But was the whale still placidly feeding? Or had their approach gallied it—frightened it away? If it was feeding, it would reappear in the general vicinity within 30 minutes or so. If it had departed, it could swim for 10 miles or more before surfacing for air. Silent

Clinging tightly to the masthead, highest perch on the ship, an alert lookout points toward a herd of whales in this fanciful 1854 American painting. In fact, a crewman would not have risked climbing to the mast top when he could just as well keep watch from his customary post on the topgallant crosstrees—a small platform three quarters of the way up the mast.

and tense with expectancy, the whalemen could only wait and see.

Although it must have been the furthest thing from young Nelson Haley's mind at the moment, he and his five shipmates sitting in their solitary whaleboat represented in microcosm one of history's most fascinating industrial adventures. Nearly 15,000 other men, in some 500 New England whaleships scattered across the oceans of the world, were variously engaged in this same perilous business: the pursuit, capture and utilization of the largest and most formidable creatures known to man.

Of some nine species of great whales—cetaceans in scientific terminology—five were regularly hunted during the heyday of 19th Century whaling and, measuring 30 to 60 feet and weighing 35 to 65 tons, any one of these immense beasts could wreak havoc among the fragile boats of the whalemen. Indeed, the whales killed dozens of their pursuers annually—and no one even bothered to count the maimed and injured.

Yet the toll was regarded as nothing compared with the gain. So enormously valuable was the whale's vast quantity of fine, light oil and many other products that whaling in the mid-1800s became a major industry that was worth close to ten million dollars annually and employed nearly 50,000 workers aside from the whalemen themselves.

With whales in sight off the port bow, crewmen of the Charles W. Morgan swiftly reduce sail and prepare to launch boats for the chase in this meticulous modern-day re-creation of the 19th Century scene by John Leavitt. Launched in 1841 in New Bedford, Massachusetts, the Morgan stayed in service as a whaler for 80 years, logging 37 long-distance voyages, more than any other whaling vessel in history.

What is more, it was an American—a New England, to be precise—monopoly. After years of lively, sometimes bitter international competition, the Yankees of New Bedford, Nantucket, Martha's Vineyard and other coastal New England towns had emerged at mid-century unchallenged around the world. New Englanders commanded and owned nearly eight out of every 10 whaleships on the high seas. Navy Lieutenant Charles Wilkes, after leading a United States expedition exploring the Pacific in the 1840s, wrote that "our whaling fleet may be said at this very date to whiten the Pacific Ocean with its canvas." In truth, the whaling voyages across the Pacific were America's greatest adventure in the western ocean, and they influenced the peoples of the area more than all the world's naval and merchant seamen combined.

The Yankee whaleman who accomplished all this was, in his way, nearly as fascinating a creature as the marvelous beast he pursued. Persevering and pragmatic, he exemplified as did no other mariner of his day the great American frontier tradition with its fierce acceptance of the challenges of the unknown. And like the frontiersman, he was shaped by his struggle, often possessed of superhuman endurance, brave beyond measure, sometimes needlessly cruel, nearly always homesick, a God-fearing man and a greedy one voracious for nature's bounty—if he could live to secure it. And when he went bobbing across the open ocean to attack the monstrous whale, his chief concern, like that of Nelt Haley at the moment, was with how great the odds were against him, and in favor of his adversary.

Haley set down the full story of that 1850s voyage—and the sperm whale as big as a mountain—in wonderful detail in *Whale Hunt*, a volume of reminiscences written many years later. As Haley had hoped and prayed, while the seamen waited on the tossing Pacific his whale was feeding, totally unaware of the men who had come to kill it. Though it would devour nearly any sort of fish, the sperm whale's favorite food was cuttlefish, octopus or squid—rubbery, many-tentacled creatures, some species of which grew to 50 feet and inhabited the ocean depths.

To find them, this whale had gone straight down—500, 1,000, 2,500 feet into a black void where his search for food depended on his hearing and his echo location system. The whale heard the cacophony of clicks, buzzes, snaps, grunts, squeals and booms of other creatures and listened for any fleeting sound emitted by potential prey. He interpreted the echoes from the underwater world around him.

At last a squid—itself huge, with 20-foot tentacles—came in range. Instantly the whale was on it, clamping the waving tentacles in its deadly jaw. Its prey thrashing desperately, the whale now turned and swam for the surface, racing up through the lightening sea, trailing a streak of phosphorescent slime from its victim. The squid lashed at its captor's square head with its great tentacles, lacerating the thin outer skin and trying to block the blowhole. But the squid was crunched into pulp by the time the whale burst through the luminescent canopy of the surface.

A split second later the whale's great lungs expelled with a violent heave. A geyser of air and moisture shot up from the blowhole at an angle of about 45° forward of the head and arced into the ocean.

Beaming proudly, Captain Nelson Haley, handsome in formal attire, holds the hand of his bride, Charlotte Brown, in this wedding day photograph taken in Honolulu in 1864. Haley learned his trade on the Charles W. Morgan, serving for four years as a harpooneer, and always remembered his first successful hit. "Never in my life," he said, "have I had such a feeling of relief and pleasure."

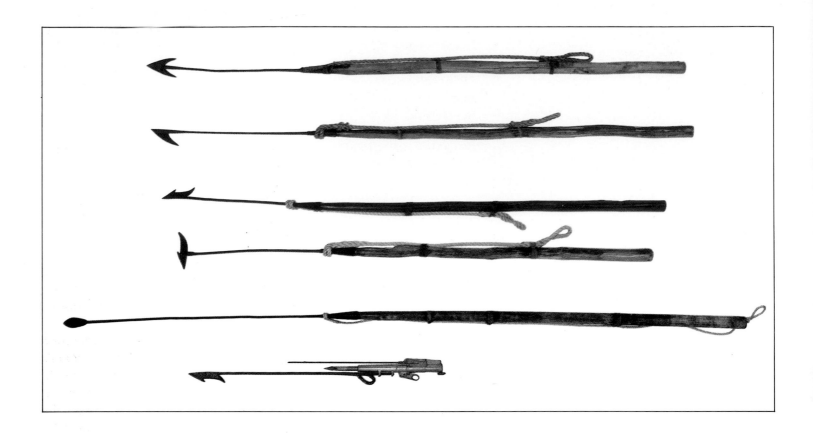

In Nelson Haley's whaleboat an oarsman cried out, "Hell and Jews-harps, there he blows!" The whale had surfaced about half a mile away. Evidently it did not hear the whaleman's outburst; at least it exhibited no alarm. In any case, it could not dive again until it had refilled its lungs. As the whale submerged slightly and rose again, spouting and drawing great breaths of fresh air, Haley and the other men in his boat heaved at their oars.

The whaleboat skimmed across the calm water, with the second mate calling softly, "Spring, boys, on your oars. Spring hard, I tell you!" As the distance closed, he ordered the oars shipped and paddles brought out. Paddles were quieter.

"It was a glorious sight," Haley recalled, "sitting on the gunwale of the boat, facing toward the whale that lay so still with his hump sticking some two or three feet above his back, which was hardly buried except when he slowly dropped his immense head and sent a swash of the sea rolling across it. He was blowing his breath from his spout-hole in such a lazy and quiet manner that the sound of it could not be heard more than a hundred yards away, calm as it was.

"Not a word was spoken above a whisper," Haley remembered, "and every man used the utmost care that the blades did not strike the side of the boat." Nine months earlier, during a pursuit of a herd of whales, one paddler had carelessly banged the bottom of his boat and the entire herd, Haley recalled, had "disappeared like so many grindstones dropped into the water."

The double flue, or barb, harpoon (1) was standard on board Yankee whalers until the 1840s, when the single flue (2) proved to cut deeper and hold better. Much more effective was the Temple toggle (3), invented in 1848, which penetrated like a needle yet held fast when the barb pivoted open (4). The lance (5) delivered the mortal thrust at close range. The darting gun of the 1860s (6) carried a charge that drove home a lethal second dart.

Not this time. The whaleboat was only about 20 feet away now, and the second mate called quietly to the bow: "Stand up." Haley gently laid his paddle on the floorboards, away from the line that led to the harpoons. Rising gingerly, he turned and took up his harpoon. It was a six-foot shaft attached to a 30-inch harpoon iron; at its end was a needle-sharp steel barb on a pivot, designed to toggle, or open so it could not pull out. Haley had meticulously sharpened his harpoons before placing them in their exact spot at the bow of the boat. As he lifted the weapon and leaned into the "clumsy cleat," a knee brace cut into a plank across the boat's bow, he studied the behemoth before him.

The whale was like a floating black island, rising and falling and rolling slightly from side to side in the gentle swells. The black hide was patterned with patches of barnacles, old scars and fresh lacerations from the giant squid on which it had fed. The circling petrels formed a halo over the whale as they pecked at the sea lice amid the barnacles and plucked fragments of squid from the water. The whale ignored them as it proceeded unhurriedly on its course at three to four knots, gathering air into its lungs for another feeding dive.

Haley and the second mate were determined not to give it time for a second lengthy disappearance. The whaleboat moved in on the whale's right side, aiming to close to within touching distance—"wood to blackskin," the whalemen called it. As they eased nearer, Haley looked at the whale's flukes, as large as the boat and undulating just beneath the surface. He could feel and smell the rank mist from the whale's sighing exhaust. He was now within striking distance, inside what the author of *Moby Dick* called "the charmed, churned circle of the hunted Sperm Whale." Still the whale was unaware. On the surface of the water its hearing was not well adapted to its surroundings, since its ears did not receive airborne sounds as acutely as they picked up waterborne sounds. And its small eyes, set in the sides of its broad head, could not see forward or aft beyond an arc of 125° on each side.

Raising his harpoon with both arms, Haley waited for the rolling back to surface once again. The huge head, one-third the length of the whale, eased under the water. The target between the head and body rose to the surface. And Haley struck.

The harpoon flashed straight down into the thick blubber and sank in clear to its socket. The black island of flesh quivered for a moment of shock. Haley quickly grabbed a second harpoon, attached to the same line, drove it up to the hilt right behind the first one and tossed out 100 feet of the line.

In the same moment the second mate shouted, "Stern all!" The long oars bent as the whaleboat backed swiftly away out of the foaming maelstrom caused by the aroused whale. Haley watched in fascination as the whale raised its huge flukes 20 feet out of the water and then brought them crashing down "with force enough," he wrote, "as to half fill the boat with water."

But Haley had little time to gape. In the next few seconds, before the enraged whale was fully under way, he had to perform the ticklish maneuver on which everything now depended. He had to exchange places with the mate, who with his greater experience would be respon-

Bent double and twisted like taffy by the gyrations of a wounded whale, this harpoon was fashioned from malleable iron, which would bend rather than break under stress. Harpoons were often mangled during the hunt, and though this one must have seemed beyond repair, most were recovered and straightened.

A tale of frustration told in stamps

Monotony was no stranger to whalemen. Nor was frustration. And nowhere were these twin conditions of life more graphically illustrated than in the whaler's daily logbook. Traditionally decorated with whale images made by wooden stamps, the log provided a visual record of the days when whales were sighted or killed. At the end of a voyage the logbook allowed the owners to judge immediately the success of a cruise that might have lasted four years.

On American ships it was the duty of the first mate to keep the log. Thus for November 21, 1838, the mate of the *William Baker* stamped his log, at right, with the picture of a whale head up, and recounted how the crewmen saw and chased some right whales, which were struck by the starboard boat and killed by the waist boat. Next to the whale the mate wrote "S.B.B. 55 bbs"—starboard-bow boat 55 barrels of oil—and added a watercolor sketch of the whale being harpooned.

The other stamps and entries tell of long days without success. A stamp of half a whale with its flukes up meant that a whale was chased but was able to escape before it could be harpooned. The head of a whale meant that a whale had been struck but escaped. And on November 24, a whole whale head down depicted the sad fact that a whale had been killed but sank before it could be secured. The creatures with long beaks were porpoises that the whalemen killed for food.

All too often, the pages of a logbook were devoid of any stamps at all in the days and weeks spent rocking along on an ocean seemingly devoid of life. In the log of another whaleship, the *Acushnet*, the mate dejectedly identified himself on October 26, 1845, as "your humble servant employ'd in killing time." And a few days later he observed that the wind was "from Oh! I don't know where, or about the same place the whales are."

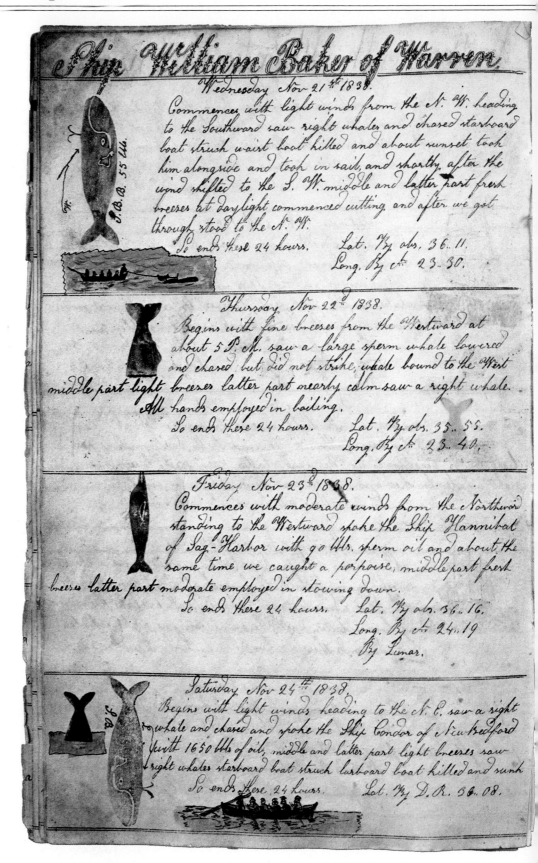

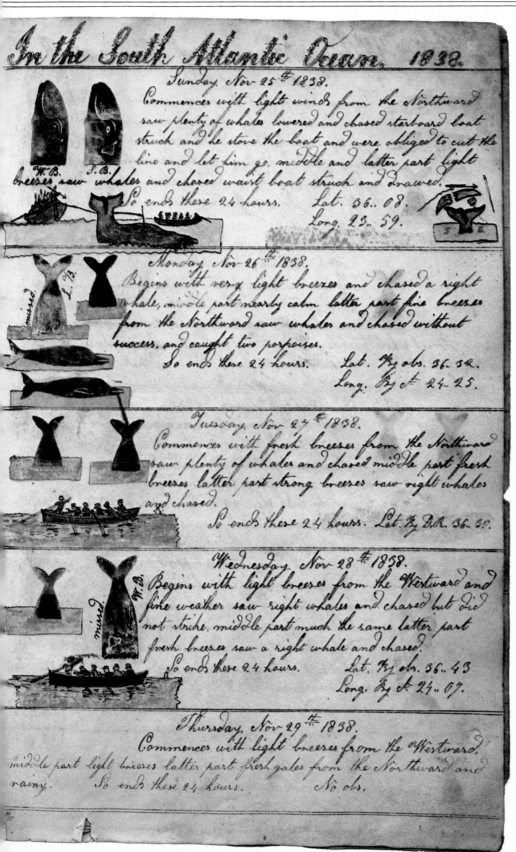

23

Two pages from the logbook of the William Baker, chronicling nine days in November 1838, are filled with a profusion of images that symbolize the success or failure of the whale hunt. The keeper of this log was more artistic than most, and added vignettes of the hunt, including one at upper right of a boat that was stove in by a harpooned whale.

Three decorative wooden whale stamps, whose inked images represent a sperm whale, a right whale and the upturned flukes of a sounding whale, typify those used on board Yankee whalers to record the killings and sightings of whales in the ship's logbook. Many whalers carried sets of five or more hand-carved whale stamps, which were fashioned from wood blocks or pieces of whalebone to resemble stylized whales, porpoises, blackfish and even turtles.

sible for lancing and killing the whale when and if they could get close enough. It would be Haley's task to handle the harpoon line and steer the boat from the stern.

Grasping the shoulders of the oarsmen, harpooneer and mate deftly stepped the length of the whaleboat. And not a moment too soon, for now the harpoon line was singing out of the boat. The wounded whale had dived again, but this time with two harpoons burning in its flesh.

The harpoon line, coiled in two tubs amidships in the whaleboat, ran not directly to the bow of the boat but first aft and around a snubbing post, called the loggerhead, at the stern and then forward over a roller guarded by a chock pin at the bow. As the whale sounded and the line raced into the water, Haley took two turns around the loggerhead and applied drag to the line. He also took up a pair of nippers, folded canvas holders, with which he grasped the line and attempted to increase the drag. The line slowed somewhat as the tension increased; Haley, in effect, was playing the whale as a fisherman would a trout or salmon.

With the line now taut, the bow of the whaleboat dipped into the water until there were only a few inches of freeboard. The second mate glanced apprehensively back from the bow and shouted, "You look out what you are about! Do not box the boat down any more—you may turn her over."

"Aye, aye, sir," Haley replied, but he let up only slightly. The line whirred faster around the loggerhead. The wood was smoking from the friction, and the nearest oarsman dipped water onto the loggerhead to keep it from catching fire, carefully keeping his balance in the process. One lurch could upset the delicate balance of the boat and tip everyone into the water—and, of course, lose the whale.

It was a tricky business for Haley both to snub the line and to mind the steering oar. He could not do both at once of course and so, while the whale was taking out line, Haley peaked the oar by slipping its handle into a leather band on the gunwale, which would keep it out of the water. The boat sped along, towed by the whale. Once the whale tired and surfaced, it would be up to Haley to use the long oar to steer the boat past the dangerous flukes to within killing range. Indeed, his future as a whaleman depended on his dexterity as a boat steerer as well as his skill as a harpooneer. If he did both jobs well he might one day win promotion to the coveted position of mate and a greater share of the kill. And who could tell? Many an excellent mate had in time risen to the zenith of ambition and had become a captain.

Haley, at any rate, was doing his difficult jobs well. His two irons not only held firm but were so deeply placed in the vulnerable area above the flipper at about eye level that they had weakened the whale. After taking only half the line from the first tub—about 900 feet— the whale slowed, and the line slackened. "He's coming up!" Haley shouted forward to the mate. "All right, let him come," called back the mate, who had already unsheathed one of his lances, 11-foot spears with razor-sharp steel oval blades.

Now the oarsmen took hold of the harpoon line and began a grunting, straining tug of war with the huge sperm whale on the other end. As the whale rose toward the surface and the men hauled themselves in its direction, Haley made sure to keep the tension around the logger-

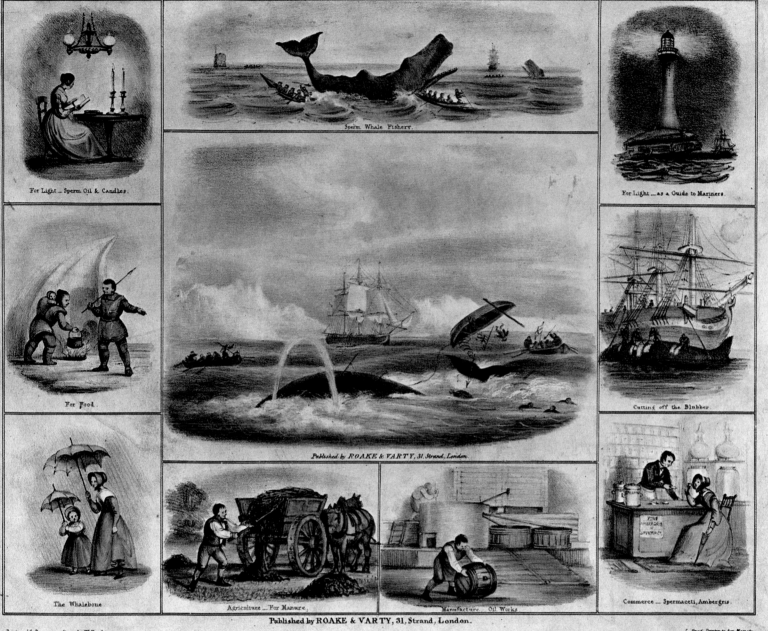

Even as two great sperm whales struggle desperately for survival in mid-ocean, their fate is already coolly anticipated in the pictures of the surrounding panels of this Victorian lithograph touting the animals' "utility to man." Whales were the source of products that were worth between eight and nine million dollars annually, ranging from a substance called baleen, used for bracing milady's umbrella, to the oil employed to illuminate homes and lighthouses.

head and took special care to coil the incoming line carefully in the bottom of the boat in case the whale should suddenly dive again or go surging off on a run.

Meanwhile the mate studied his target: it was an area about four feet by three feet square where the animal's massive arteries clustered near its heart and lungs. Whalemen called it the "life" of the whale. It was critical to strike this point squarely; a miss would further enrage the whale, and in these close quarters, that could be fatal to the whalemen. But the aim was true. The lance sliced with terrible efficiency into the pulsing junction of arteries and penetrated the windpipe. Almost immediately the whale's lungs were flooded with blood and the supply to the heart dwindled.

"Stern all!" the mate called again. The boat backed off. The whale submerged, surfaced and spouted a pinkish-red mist into the air. "We have him now," said the mate. And one of the oarsmen cried, "See! See! His chimney is afire!"

The whale's death throes were mercifully brief. After about five minutes it rolled dead on its side. Blood stained the sea. The petrels danced over the water, wings fluttering like butterflies, as they fed on the blood.

The chase and capture had taken a little more than two hours. The afternoon sun had set. With dusk a breeze came up, and the men in Haley's boat were spared a grueling tow to the ship, which now could run down to them. While they waited, they rinsed their hands, blistered and bleeding from the oars and the harpoon line, and lighted the lantern in their boat to guide the ship to them. Then they cut a hole in the whale's flukes and passed a doubled rope through it. When the ship hove to, the rope was replaced with chain and the heavy carcass, floating just awash, was hauled alongside, flukes forward.

Haley's boat went rattling up to its davits, and the young harpooneer jumped onto the *Morgan's* deck to join his shipmates, who were leaning over the rail admiring his prize. The whale measured 60 feet from jaw to flukes, by far the biggest one that had been brought alongside the ship so far on the voyage.

But Haley's euphoria swiftly vanished when an officer called out that the captain wanted him and the second mate to report to his cabin. In the excitement, Haley had completely forgotten that he and the mate had committed a minor act of insubordination by their failure to follow the captain's whaleboat. As it turned out, he had been right and the captain wrong. But he well knew what a stickler for the rules Captain John Samson could be.

"I timidly crawled down the cabin stairs," Haley recalled. He paused a moment to listen at the door—"to hear a word or two that would tell me what kind of humor the Old Man was in."

What Haley heard was the Old Man laughing. "With my hat in my hand," Haley wrote, "I boldly stepped into the cabin."

"Nelt," said Captain Samson, "come here. You have covered yourself with Glory today, and must take a drink with me." Haley claimed that he was not a drinker. However, on this occasion the youngster poured himself what he regarded as "a good tot of liquor. Raising it to my lips, at

Displaying the burgeoning importance of New Bedford, Massachusetts, as a whaling port, this decorative 1857 broadside identifies the private ensigns and the ships of the town's 100 whaling merchants. Whaling was a family concern: many names appear more than once in the agents' listing, and no fewer than eight Howlands controlled some 50 ships.

the same time saying many returns of the day, I drank it off, amid the laughter that seemed brought about by my remark. Making a bow to him and the officers, I went on deck feeling much better in my mind than on going below.''

Not only that, but Captain Samson proceeded to break another rule of whaler etiquette. To celebrate their biggest catch, the captain invited Haley and the other harpooneers to dine with him and the officers that evening. Normally the *Morgan's* off-watch harpooneers ate at the same table as the officers but only after their superiors had been served. On this fortunate night the officers and harpooneers joined in a convivial supper of salt meat, bread and pie, with tea to wash it down.

While the exhausted oarsmen of the *Morgan* caught a few hours of sleep that night, other crewmen readied the gear for the next two days' work: rendering the giant carcass into barrels of oil, which, at the prevailing price, would bring $40 per barrel. The narrow scaffolds, called the cutting stage, were lowered over the whale on the *Morgan's* starboard side. The 20-foot cutting spades were taken from their racks over the deck. The huge iron kettles, each with a capacity of 250 gallons, were uncovered and cleaned. Kindling was laid under them, to be lighted when the blubber was ready to be boiled; and water was poured into the "duck pen," a brick trough on the floor under the tryworks, to protect the wooden deck from the intense heat of the fire.

By the first light of dawn the *Morgan* was ready to try out her whale, and the men were called into action with the booming cry from the first mate: "All hands ahoy! Tumble up and man the windlass!"

The *Morgan* carried only enough sail to keep her from drifting in circles. In the gray early-morning light the whale carcass was now chained alongside by its head as well as its flukes. The sharks had arrived in the night and were feasting. The men fell to their tasks with a will, and for the next 50 hours, working in shifts, amid an inferno of smoke and flame and oily stench, they flensed, sliced and boiled the whale's thick blanket of blubber into almost 2,000 gallons of the finest sperm oil. In addition, there was ivory from the whale's great teeth and a second oil known as spermaceti, which came from the whale's huge head and was highly esteemed by New England candlemakers.

Aboard many whalers the trying out of a new whale was an occasion that called for a tot of rum for everyone. But Captain Samson, like many another New England skipper, was a puritanical man; even though he had hoisted a glass with Haley, drinking to him was almost as sinful as card playing—which was absolutely forbidden on board his ship. There would be no general ration of rum. After a short break for a rest and a bit of food, the crewmen set about the messy task of cleaning up. The thick scum of fat was mopped from the deck; the blubber room and the try-pots were scoured with lye from the furnace ashes; the grease and soot were washed from the yardarms. The men's clothes were first soaked in a mixture of salt water and urine, which had been collected in a barrel for its grease-cutting ammonia content, and then were rinsed in salt water.

Some merchant sailors claimed that they could smell a whaleship miles away. In *Two Years before the Mast* Richard Henry Dana recollected anchoring near a whaler and finding both ship and crew slovenly and stinking of oil. But Nelson Haley wrote that aboard the *Morgan*, after each trying out, Captain Samson insisted on having the ship scrubbed down so that "when all was done you would have found it hard work to soil a white pocket handkerchief by rubbing it on any part."

For the next three days both rails of the ship were lined with casks while the oil cooled. Because the staves and hoops contracted as it cooled, the ship's cooper had to make constant adjustments to keep the containers from leaking—a task that would have been impossible had all the casks been piled in layers belowdecks. But at last the casks were stabilized and were manhandled into the hold by the exhausted crew. As for the whale's enormous carcass, it had long since been cast adrift to feed the sharks.

The *Morgan*'s course was laid south toward the coast of New Zealand. Her men once more assumed their perches in the crosstrees, peering out over the shimmering horizon in search of the next telltale, angular spray that would bring the cry, "Thar she blows!" Then there would be another dangerous chase and another cutting in and trying out—day after day, month after month.

The mammal that returned to the sea

From the earliest days of whaling, men speculated about whether the beast they pursued was a fish with certain mammalian characteristics or a mammal masquerading as a fish. The whale was warm-blooded, suckled its young and breathed through lungs. But so far as most whalemen were concerned, the weight of evidence clearly indicated fish. As one whaleman argued: "Whales has slick skins, so has fish; whales has fins, so has fish; whales has tails, so has fish. I conclude then, whales *is* fish."

It was not until the mid-19th Century that naturalists began to unravel the puzzle. Comparing fossil skeletons of early whales with those of terrestrial mammals, naturalists determined that whales—and their cousins the porpoises—were, in fact, descended from a group of hippopotamus-like mammals that long ago inhabited the earth.

While some of these mammals evolved into animals such as cattle, goats, horses and rhinoceroses, another group returned to the sea 60 to 65 million years ago in search of food. Over the eons, forelegs developed into flippers, hind legs gradually disappeared, the tail broadened into fan-shaped flukes, and the mammal's hairy coat was replaced by a thick layer of insulating blubber.

About 85 species of cetaceans, as scientists call these creatures, survive today, ranging from porpoises scarcely a yard long to whales measuring 100 feet. Naturally, it was to the largest members of the order that man turned his attention. Of the nine species of so-called great whales, the five shown on these pages formed the basis of 19th Century whaling. Indeed to the whaleman, all other animals, as a naturalist put it, were "contemptible in comparison."

Averaging 55 feet in length and 63 tons in weight, the male sperm whale was nearly twice the size of its mate and the largest of the toothed whales. Because of its seemingly unlimited numbers (possibly 1.5 million worldwide in the early 1800s) and because of its great value (up to 1,890 gallons of oil from one animal), the sperm whale was the species on which the great age of Yankee whaling depended.

Its scooplike lips curved back as if in a grotesque smile, the 50-foot, 42-ton bowhead displays the flexible comblike baleen that could grow to lengths of 14 feet. This arctic whale had not only the longest baleen of any cetacean, but blubber sometimes more than two feet thick. Since the 3,000-pound yield of bone, selling in the late 1800s at three dollars or more per pound, was so valuable, whalers sometimes jettisoned a bowhead carcass without even processing the blubber.

Except for its barnacle-encrusted bonnet—a large callus on its head that served no known purpose—the right whale was similar in size and appearance to the arctic bowhead, but inhabited more temperate waters in both the Northern and Southern Hemispheres. It got its name as the right whale to hunt because it was a lethargic swimmer, floated when killed, and possessed blubber that was 16 inches thick and baleen that grew up to eight feet in length.

Matternes

Averaging 47 feet in length and some 40 tons in weight, the relatively slender gray was one of the smaller whales, but was still almost 20 feet longer than the whaleboats (above) of the Yankees who pursued it along the Pacific coast. It was the only whale of the baleen species to feed on bottom-dwelling creatures — clams, crabs and marine worms —and frequently surfaced after a meal, said one captain, with "head and lips besmeared with dark ooze from the depths below."

Although the 45-foot, 35-ton humpback had thick blubber and baleen up to 27 inches long, this widely distributed coastal species tended to sink when dead, making it the last choice of 19th Century whalemen. But they found the humpbacks fascinating. Observers swore that these creatures turned backward somersaults in mid-air. And one whaleman wrote that humpbacks used their long flippers "to administer love pats which may be heard at a distance of miles."

It was May 27, 1853, more than two years later, when the *Charles W. Morgan* was finally sighted from the rooftops of New Bedford, Massachusetts. Under full sail before a fresh northwesterly, she came boiling up to Buzzards Bay, hauling her mainsail aback only long enough to take aboard her pilot. The breeze obligingly backed into the southwest, and the *Morgan* raced in between Clark's Point Light and Sow and Pigs rocks. She looked as if she had just left home. Her hull had been repainted in New Zealand before she headed for New Bedford, and on the way up the Atlantic her tattered rigging had been replaced. The only sign of her arduous voyage was the gap on the foredeck where her tryworks had been. The oven had been torn down brick by brick and tossed overboard, and her two huge iron try-pots were now lashed on deck upside down.

It was low tide as the *Morgan* glided smartly past Palmer's Island, her sails coming down on the run. Captain Samson drove her straight into the mud 90 yards from the wharf. One of her whaleboats, with Harpooneer Nelson Haley in it, was dropped into the water to take ashore the lines with which the ship would be warped alongside the wharf with the rising tide. When Haley tossed a line onto the dock, it was caught by a waiting cousin.

The young harpooneer was followed up the wharf by a worshipful crowd of youngsters. "The boys surrounded me three deep," Haley remembered, and escorted him to his first visits ashore: a barbershop for a haircut and a bath, and a clothing store for a suit that would fit properly and "not require reefing in back, arms or legs," as did the gear he had procured from the ship's slop chest. Similar groups of friends and relatives were meanwhile swarming over the wharf to welcome the other members of the *Morgan*'s crew.

They had brought back 1,150 barrels of oil, each containing about 30 gallons, and Captain John Samson had also loaded 1,000 barrels for another whaler, plus a cash cargo of two or three tons of kauri gum, a resin found in New Zealand conifers and brought to the United States to be used in varnish. While waiting for the *Morgan*'s whale oil to be tested and sold, and her accounts to be settled, young Haley took "the cars," as he put it, to Portland, Maine, where "having Mother's arms around me, all sense of fear left that had hovered over me, that something might occur to prevent our ever meeting again." A week later he was back in New Bedford presenting himself at Edward M. Robinson's countinghouse to be paid off.

The *Morgan* had grossed $44,138.75 on her voyage. Under the normal rules for division, a third of this went to the owners, another third for expenses of refitting the ship for her next voyage, and a third to the officers and crew. Thus Robinson and his investors divided a bit more than $14,000 and Captain Samson probably received about $4,000— which was, in those days, a decent if not a princely sum.

Harpooneer Haley's share was $400. In the course of his voyage, Haley, like his shipmates, had asked for cash advances and had made some purchases from the ship's slop chest. Haley's advances and purchases added up to $200, leaving him a total of $200 for four years of danger and drudgery. It was, Haley said, "a rather slow way to get rich."

Modern whaling: the age of mechanized slaughter

The Yankee whalemen of the 19th Century scarcely paused to reflect on the damage they might be doing to the world's population of great whales. They acted in the Victorian certainty that all the creatures of the earth existed solely to serve man in one fashion or another. And in their single-minded pursuit of the whales, they visited calamity on some species in some of the world's whaling grounds.

In the North Atlantic, for example, the once-large herds of right whales, a slow and vulnerable species, were reduced to uneconomic proportions by the early 1800s. The whalemen next concentrated on the sperm whale, taking great numbers of this huge animal until it, too, was hard to find in the Atlantic. The whale hunters then moved into the Pacific and Indian Oceans, where they discovered so many sperm, right, humpback and, later, bowhead whales that no Yankee whaleman could imagine the stocks ever being exhausted.

There were more than a million other whales that the early whalemen could not hunt at all. Most of the rorqual group—among them the enormous blue whale, the fin whale and the sei whale—were too big and too fast for men with hand-held harpoons to attack successfully. And on those infrequent occasions when rorquals were killed, they almost always lost their buoyancy and quickly sank—to the whalemen's rage and frustration.

But the advent of modern European whaling changed all that. With advanced technology—cannon-fired explosive harpoons, fast steam-powered hunter boats that pumped compressed air into whale carcasses to keep them afloat, enormous factory vessels that could, by 1975, process the oil and flesh of dozens of whales each day—the assault on the whales reached unbelievable intensity.

Between 1842 and 1846, at the pinnacle of their success, New England whalemen returned home with the oil of some 20,000 sperm whales in their holds. Between 1960 and 1964, whaling fleets—mostly Japanese and Russian—killed 127,000 sperm whales.

What is more, 20th Century whalemen were not limited to the traditional quarry; with their sophisticated weapons they could kill any whale that swam. The rorquals became easy prey.

The huge blue whale, greatest of them all—it was called the "cream" whale because one carcass produced two to six times the oil and flesh of any other whale—was hunted so assiduously that almost 30,000 were taken in the year 1931 alone. Between 1910 and 1967 a total of 330,000 were killed and their bodies processed into cosmetics, lubricants, auto transmission fluid, pet food, cattle feed, fertilizer and margarine; and a considerable amount of whale meat was consumed by humans. By 1967, out of a population of blue whales believed to have once numbered between 400,000 and 500,000, only about 14,000 remained.

In the same way the fin whale, once thought to have numbered 400,000, was reduced by some estimates to one-fifth that total; the sei whale was killed off to the point where only half of the original 220,000 existed.

By careful estimates a grand total of more than 1.5 million whales of all types were killed in the half century ending in 1975. In the 1930s the right, bowhead and gray whales were belatedly accorded full protection by the International Whaling Commission, as later were the blue and the humpback. Other species, such as the fin, sei and sperm whales, were placed under a protective quota system. Nevertheless, as late as 1975, in excess of 17,000 sperm whales were being killed annually, four times more than the old Yankee whalemen took in 1846, the greatest year in their history.

Triumph's hellish ritual: cutting in and trying out

After all the peril and excitement of the chase and capture, there came a ritual of triumph that whalemen, as an observer named J. Ross Browne wrote in 1846, approached with "feelings of mingled disgust and awe." This was the business of "cutting in" and "trying out"—butchering the enormous beast and rendering its blubber into oil.

It took two to three days to process a single whale. If several had been caught, the whaler was transformed into a miniature factory afloat for five, sometimes six days. The crew worked around the clock in six-hour shifts, stripping the carcasses and feeding the huge try-pots. The captain kept a sharp eye on everything, making sure that the men cut the blubber to the proper thickness so that it would render the most oil, that they frequently stirred the cauldrons to prevent the blubber from settling to the bottom, that the fires were fueled to the highest intensity.

During trying out there were no regular meals. Now and again the men would take a break from their furious work for a smoke and a bit of food, dipping biscuits in salt water and frying them in the bubbling oil; sometimes they would mince whale meat and mix it with potatoes into a sort of fritter that, recalled whaleman Browne, "answers very well for variety."

As the work progressed, dense clouds of smoke obscured the rigging, and the decks gleamed with blood and slippery grease. Oil hissed and sizzled in the try-pots, and the ship was enveloped in a stench that would have been unbearable to a landsman.

Even to a veteran whaleman, wrote Browne, "a trying-out scene has something peculiarly wild and savage in it. There is a murderous appearance about the decks and the huge masses of flesh and blubber lying here and there, and a ferocity in the looks of the men heightened by the fierce red glare of the fires. I know of nothing to which this part of the whaling business can be more appropriately compared than to Dante's pictures of the infernal regions."

A slain sperm whale has been brought home to the starboard side of the ship, with its head pointing aft, and crewmen have secured it by running a heavy chain around the flukes. A platform known as the cutting stage is being lowered into position above the whale. Holding razor-sharp 20-foot cutting spades, the flensers—the men who will strip away the blubber—stand ready on deck while other crewmen break out their tools: gaffs, pikes, blubber hooks and forks. With the whale's tail pointing forward, the whaleship will maintain a slight headway under light sails during the entire cutting-in process so that the forward motion will act to hold the huge carcass in close to the hull.

As sharks swarm around, a harpooneer with a "monkey-rope" life line around his waist has jumped onto the carcass. He is inserting a 200-pound blubber hook into a hole cut between the whale's eye and a flipper in order to hoist a huge "blanket piece" of blubber that is being cut by the flensers. It will take a dozen crewmen at the windlass to lift each 2,000-pound piece, 15 feet long by 3 feet wide, onto the deck.

Positioning themselves on the cutting stage, the mates wield their spades with surgical accuracy to sever the tooth-studded lower jawbone from the whale, which has now been rolled onto its back. Next they will sever the massive vertebrae and decapitate the whale, after which the head, weighing perhaps 16 tons, will be moved aft to wait while the men finish attending to the blubber.

As the last great blanket piece is lifted to the deck, flensers probe deep with their spades, seeking ambergris, a solid substance that sometimes collected around indigestible matter in the intestines of the whale and was worth far more than its weight in gold. Highly prized for its properties as a fixative in perfumes, ambergris was so rare that from 1836 to 1880 the entire American whaling fleet found less than a ton of the stuff.

Having skewered the Bible leaves with a two-pronged blubber fork, a crew member heaves the blubber into a try-pot set in the heavy brick furnace, or tryworks. A single try-pot could render 200 gallons of oil from blubber in about an hour.

With the head hoisted up to deck level near the gangway, a whaleman carefully lowers a bucket into the "case," a natural reservoir containing hundreds of gallons of an oil called spermaceti. So fine was this oil that it required only a quick scalding to prevent spoilage before being poured directly into casks.

On his knees amid the blubber, a crewman belowdecks cuts a large blanket piece into smaller chunks called "horse pieces." A second whaleman, wielding a keen mincing knife, cuts the horse pieces into "Bible leaves," thin slices, resembling the pages of a book, that melted rapidly in the try-pots.

Their faces illuminated throughout the
night by the fires of the tryworks, crewmen
tend the boiling cauldrons filled with
Bible leaves. At left, a whaleman is ladling
the rendered oil into a cooling tank,
where it will rest until cooled before being
placed in casks. At right, another man
skims shriveled, spent Bible leaves
from a try-pot; these will now be used
instead of wood to keep the fires blazing.

Their clothes thoroughly soaked in
oil, weary sailors belowdecks wrestle a full
cask into stowage. Oil casks were
ingeniously designed in a variety of sizes to
fit into every nook and cranny in the
hold. After trying out, whale oil could be
stored almost indefinitely; many
cargoes arrived in port without spoilage
after as long as four years on board.

Chapter 2

A heroic tradition from Stone Age times

*An inbound whaler, her sails clewed up
in the brisk breeze, steers across the mouth
of Massachusetts' Acushnet River,
where lay the ports of New Bedford and
Fairhaven. In 1854, when this picture
was painted by William Bradford, these
two New England towns sent forth
a whaling fleet of 359 ships—more whalers
than all other American ports combined.*

ompared with other young Yankees growing up quietly in the rustic New England of the 1850s, Harpooneer Nelson Haley was a bold adventurer—the envy of all his peers crowding the New Bedford wharves. But Haley was following a heroic tradition far older than he could possibly have imagined. Whaling extended back into the dim reaches of history, when Stone Age men had braved the terrors of the unknown sea to challenge the great air-breathing monsters. The flesh of one whale might keep an entire community alive through the most brutal of winters, and harpoons fashioned from bone have been found in paleolithic cave sites along western European shores. In Scandinavia, rock carvings dating to 10,000 B.C. depict crude harpoons, skin boats, and whales as well. And in the land of the Basques, around the perimeter of the Bay of Biscay, whaling very early progressed from random, primitive slaughter into an organized hunt.

Whaling had been a part of Basque life from the late Stone Age, when the Basques first appeared on the northwest coast of Spain and across the Pyrenees on the extreme southwest coast of France. Where the Basques originated is unknown, but they were a people entirely different in language and custom from their French and Spanish neighbors. In the 17th Century a French cleric commented acidly: "The Basques speak among themselves in a tongue that they say they understand, but I frankly do not believe it." Nevertheless, a few of their curious words did find their way into other languages. One of them was *arpoi*, meaning "to grasp or to hold," which in time became the Spanish word *arpon*, from which came the English harpoon.

The first Basque whale hunters probably attacked whales stranded in the shallows. But by 700 A.D., according to ancient texts, the Basques were paddling out into the Bay of Biscay in flotillas of small boats and mounting an organized assault against the beasts they called *sarda*.

Every autumn the Basques kept watch from the moist verdant hills of Spain's Biscayan coast for the great congregations of migrating whales to come rolling through the Bay of Biscay. Upon the first sighting, men stationed atop stone watchtowers along the coast would set fire to bundles of damp straw as signal smoke, and beat small drums. Below on the brown sand beaches, other men would come running and launch their boats—as many as 10 oarsmen in each, as well as a harpooneer and a steering oarsman. The ensuing chase and battle was an act of extraordinary courage for medieval man, who for the most part regarded the sea with terror.

"Each year many die in these battles," marveled the Venetian historian Andrea Navagero, "because of the resistance that the beast opposes. When they discover that a whale is heading for the land, well-manned boats are put out in great number and cut off its retreat to the open sea. When the whale comes to the surface to breathe, those who are in the boats fling small tridents fastened to cords. The monster, feeling the blow, makes a great to-do, rushing toward the boats and striking them with his tail.

"Many times wounded, and finally exhausted and bound with an infinite number of ropes, which leave it no freedom of movement, the whale cannot avoid being brought in, and ultimately finds itself in such shallow water that the men are bold enough to approach and finish it off. They divide the booty into numerous parts; some of it is sold fresh, and they say that the meat is excellent; and some is salted. In fact, they draw so much meat from the beast that the whole of France could eat from one whale."

A broad exaggeration, perhaps, but there is no inflating the bravery of the Basques—or their sense of commercial opportunity. The whales they hunted and killed were not the toothed sperm whales found by Nelson Haley and other latter-day whalemen, but the right whales, among the species known as baleen whales, which possessed a remarkable feeding apparatus. Instead of having the lower jaw of the sperm whale with its great teeth, these whales were endowed with wide rows of slats, called baleen, that were composed of keratin—the same substance as that of human fingernails—ending in a fringe of bristles that served as strainers. The whales scooped tons of water into their capacious mouths, closed their huge lips and pushed against them with their massive tongues, filtering out a type of plankton or krill, the tiny floating crustaceans that constituted their diet.

The Basques not only made use of the whale's meat, but rendered its thick coat of blubber in tryworks on the beach for oil with which to light their homes. They discovered that the whale's heavy bones could be fashioned into a variety of implements, such as knives and spades. And the baleen, tough and yet flexible, found a use in whips for horses and other animals, as well as for archers' bows and soldiers' shields. When shredded and colored, baleen also made impressive plumes for warriors' helmets.

All in all, the whale quickly came to be regarded as such a valuable beast that early medieval English law proclaimed it a "royal fish," and thus property of the king. Whale's tongue was accepted as a tithe by the Diocese of Bayonne, France, whose canons enjoyed it as a delicacy at the refectory table. And in the 1400s, France's King Louis XII exacted royal taxes from the proceeds of each whale slain.

Inevitably, as the uses and value of whales increased, the animals were driven from the Bay of Biscay. By the 15th Century the Basques were ranging farther and farther out to sea, first in their sail-and-oar-powered galleys modeled on Roman designs, and later in pure sailing ships— possibly carracks as much as 60 feet long and 30 feet wide. The ship's complement numbered nearly 50 men—including a pilot, usually a neighboring Norman of Viking ancestry familiar with Norse navigation across the chill North Atlantic. Subsisting on dried beef and biscuits,

Jonah, Israel's reluctant prophet, kicks his heels heavenward as he disappears into the maw of a monstrous whale in this 14th Century medallion, painted by Giotto as part of a magnificent series of frescoes for the Arena Chapel in Padua.

A large bull sperm whale lies stranded on the beach at Scheveningen, Holland, in this 1598 engraving, while local inhabitants crowd around, exhibiting awe and industry. An axman atop the whale chops into its thick blubber, and companions below collect buckets of oil seeping from the animal. At left a group of men starts to measure the carcass by stretching a line forward from the tail.

with barrels of cider to replace their water when it grew foul, the Basques pursued the whales up the Atlantic to the Faroes, and beyond to the arctic seas around Spitzbergen. Indeed, there is evidence in early maps that the Basque whalemen of the Middle Ages probably sailed past Iceland to Greenland and Newfoundland, reaching the New World after the Vikings but a century before Columbus.

The Basque carrack was succeeded in time by the faster caravel, carrying two whaleboats to launch over the side when whales were sighted. These earliest whaleboats had flat sterns and were not as seaworthy as the later double-ended Yankee whaleboats. The whale was harpooned, lanced and towed back to the ship, where its blubber was cut into chunks and stored in casks to be brought home. The weather in those northern latitudes ordinarily was cool enough to keep the blubber from spoiling, but by the time the whaleships were approaching the Bay of Biscay the stench of rotting whale fat was enough to gag even the toughest of crewmen.

For three centuries, Basque ships sailed throughout the North Atlantic, often in small fleets, their whaleboats converging on vast herds of migrating right whales. There is no record of how many whales they killed, or of how much oil they succeeded in rendering in the tryworks they set up on shore. But that whaling was a booming industry is certain from the fierce competition the Basques attracted, commencing in the early 17th Century.

Freed by revolutionary Protestantism from their absolute allegiance to feudal rulers and from the Catholic Church's economic domination, purse-wise Protestant entrepreneurs in England and Holland were engaged in a mounting commercial and naval rivalry for domination of the world's rich sea-lanes. The British and Dutch, previously good customers for Basque whale oil and baleen, now decided with mercenary zeal to go after their own. They started by enlisting the Basques themselves as harpooneers and flensers.

By the late 17th Century, the Dutch alone had nearly 200 whalers combing the northern waters. Almost every crew included Basques whose special knowledge was rapidly being absorbed by their foreign employers. Meanwhile, the fabric of the Basques' ill-defined nation was being torn apart by the unending wars between Spain and France, both of which forced many Basques to serve in their military forces. Inevitably, Basque whaling suffered a severe decline.

The British and Dutch students soon mastered the art of whaling and sent their teachers away. Asserting their control of both the industry and the sea, the two nations even banned the Basques from northern waters, threatening to sink their vessels if they dared venture too far north or west of the Bay of Biscay. By the mid-17th Century the Basques' dominance of whaling had come to an end, and the British and Dutch were fighting between themselves to gain control of the northern whaling grounds. The Dutch finally drove the British from the North Atlantic whaling grounds.

In the late 17th Century hundreds of whalers captured thousands of whales. But the rapacity of the whalemen had a serious effect on the population of right whales in polar waters around eastern Greenland. In the arctic latitudes, summer concentrations of plankton had long brought baleen whales to congregate and feed on the Greenland shelf, before commencing their annual 3,000-mile migration southward to the equatorial Atlantic where their young were calved. But the aggressiveness of the English and Dutch whalemen destroyed this arctic sanctuary. Instead of being hunted only during their seasonal migrations southward along the European coastal shallows, the right whales were pursued all the way to the eastern arctic feeding grounds and were hunted through the spring, summer and fall. The one Atlantic feeding and breeding sanctuary remaining was in the far western part of the ocean, in Baffin Bay, the Labrador Sea and along the sparsely populated coasts to the south. And whalemen in growing numbers would soon make their appearance in those waters.

On the far side of the world at the end of the 17th Century, the Japanese were pursuing Pacific humpback whales, right whales and a few sperm whales that migrated along Japan's coast. Japanese whaling methods were quite unlike those of the Europeans. The Japanese used harpoons, but with inflated bladders attached to tire the diving whale, mark its location and save any harpoons that missed their mark. The hunters went out in flotillas of pine longboats rowed by 30 to 40 oarsmen. And they added another invention of their own: an enormous net supported by floating barrels. The Japanese technique required no fewer than 30 boats, some to round up the whale and others to handle the nets. Once the whale was trapped within the nets, it was repeatedly speared until a man could, with impunity, climb onto the beast and attach a tow line to drag it ashore.

At the same time Eskimos in the Arctic pursued another baleen whale, the bowhead, attacking it with harpoons tipped with walrus-tusk blades. Like the Japanese, the Eskimos fastened bladder buoys to their harpoons, and after teams of six men in skin boats had harpooned and lanced a bowhead it would be towed ashore for the cutting in.

To guarantee a kill, the inhabitants of the Aleutians often dipped their harpoons in a poison, possibly a derivative of aconite or the plant monkshood, that acted as a sedative to lull the whale until it could be killed with spears.

Coastal American Indians did not engage in whaling on a sustained basis. They were generally content to make use of dead whales that washed onto the beach, although they might attack a sick or injured whale in the shallows, or one that had become trapped by the tide inside a narrow inlet. Captain George Waymouth, an English explorer, reported that he observed Indians whaling along the coast of Maine in 1620. Waymouth's journal noted that the Indians went after a whale "with a Multitude of their Boats; and strike him with a Bone made in fashion of a harping iron fastened to a rope; which they make great and strong of Bark of Trees, which they veer out after him; then all their Boats come about him as he riseth above Water, with their Arrows they shoot him to death." The whale Captain Waymouth saw must have been a very small one, possibly a species known as the pilot whale, for a large whale could not be killed by mere arrows.

But none of this depredation, not the wide-scale hunting by the Basques, British and Dutch, not the attacks by the Japanese and North Pacific peoples and not the occasional success of American Indians, could compare with the harvest of great whales that would be reaped from the ocean by the Yankees of New England. New Englanders raised whaling—at least in the old-fashioned meaning of the word—to its zenith, turning it into a major industry that involved hundreds of ships cruising the world over. In time, whale oil lighted cities and homes and lubricated machines, and the tough, flexible baleen found scores of new uses—in everything from chair springs and hairbrush bristles to buggy whips, skirt hoops and corset stays (which mercifully replaced the previously used strips of iron).

The American settlers of Easthampton, Long Island, were whaling as early as 1649, and lookout towers—much like those used by the Basques—dotted the southern shore of the island. The township selectmen divided the beach front into four sectors, and each was assigned a whale patrol of 11 men. A stranded whale was worth a reward of five shillings to the finder (no reward was offered for a whale found on

Sunday—that clearly being an expression of God's providence). Schools closed during the whales' seasonal migrations from December to April, and the cry "Whale off!" stirred the whole community into action. By 1687 seven separate whaling groups were operating in the area, and throughout early spring the sky was dark with smudge from the tryworks along the shore.

Whaling followed the coast just as the whale did. As early as 1692 Cape Cod men were taking whales that had washed up on their shores, and they were pursuing in boats those whales sighted in coastal waters. In some instances these animals had been harpooned but then lost by other New Englanders at sea. Shortly after, the settlers on Nantucket took up whaling as a serious business and hired a mainlander, Ichabod Paddock, to teach them how to go to sea after their quarry. The islanders had killed a few large whales before, when the beasts had blundered into the confines of the harbor and men in boats had slain them with harpoons hastily fashioned by the village blacksmith. Now the Nantucketers found that their island, lying less than 30 miles at sea, was within easy reach of the main migration stream of right whales passing just south of the island's shoals. The carcasses were towed ashore to be tried out, and there were scores of them each spring and autumn. So skillful did the islanders become that Obed Macy later claimed in his *History of Nantucket* that in 70 years of

A harpooned whale smashes one whaleboat and tows another at a terrific rate in this primitive but dramatic depiction by an unidentified 19th Century artist—possibly a whaleman—working in oil on sailcloth. In what was known as a "Nantucket sleigh ride," a whaleboat might be dragged for many miles at speeds up to 12 knots before the prey tired.

shore whaling from 1690 to 1760 not a single whaleman lost his life.

As on Long Island and around the Bay of Biscay, watchtowers rose on Nantucket's south shore. Boat crews were formed, some camping on the beach during the season. Before long, whaling was such an industry that much of the island's life and economy revolved around it.

"In the year 1690," wrote Obed Macy, "some persons were on a high hill observing the whales spouting and sporting with each other, when one observed: There—pointing to the sea—is a green pasture where our children's grand-children will go for bread."

But naturally after a few decades of this efficient method of hunting, there was a growing scarcity of whales within easy reach. So the Nantucketers ventured farther out to sea, in small sloops trailing one or two crude whaleboats. And it was in one of these sloops that an islander named Christopher Hussey and a few stalwarts found themselves caught up in a wild storm one autumn day in 1712. Hussey and his companions were driven farther out into the Atlantic than any Nantucket whaleman before them had ever been—and when they returned, Yankee whaling was changed forever.

In the embrace of the storm, "Kit" Hussey, as he was called, ran for many hours (perhaps for days; the exact record is lost to history) before the winds and tremendous seas. At last the pelting rain began to lessen, visibility improved and Hussey saw black shapes suddenly materialize in the tall seas around him. The men were in the midst of a great herd of whales, not the familiar baleen right whales with their characteristic double spout, but enormous beasts with a low, single spout arcing forward from the vertical.

Hussey sent his sloop racing down among the Leviathans. Somehow he succeeded in harpooning one of the beasts and then managed with incredible skill—and luck—to subdue the great whale after a chaotic battle. In time he had the whale securely lashed to the side of the sloop. By now, the cyclic storm had moved out to sea so far that its winds had hauled around to the east and Hussey's sloop—and the whale lashed to the side—was being blown back toward Nantucket. The strong current running through the eastern shoals did the rest. By the time the storm had lifted, Sankaty Head at the island's eastern tip was in sight, and Hussey was home with his prize.

What Kit Hussey had encountered in mid-storm was a herd of sperm whales, each one nearly the length of his 70-foot sloop. The Nantucketers were not entirely unfamiliar with the sperm whale. One had washed up on the island's southwest shore some years earlier, and the islanders had been agreeably surprised at the richness of its blubber. It produced an oil far superior to that of the right whale they had been chasing along their shores. They had assumed, however, that this odd type of whale was extremely rare, if not some type of mutation. Hussey's discovery proved that the sperm whale was not rare; it was simply pelagic—a deepwater dweller.

Within weeks, Hussey was pushing farther and farther into the Atlantic in search of more sperm whales. Other Nantucket whalemen were soon to follow.

Shooting its enormous bulk headlong out of the water and literally standing on its tail, a right whale, curiously marked with a rectangle of white on its back, towers over its human tormentors before crashing back down into the ocean. The scene, painted in 1843 by a Delaware whaleman named John Martin, was scarcely an exaggeration; aroused whales were known to clear the water entirely.

As these pioneer Yankee whalemen quickly discovered, the sperm whale was not only a richer prize than the right whale but an even more fascinating and fearsome quarry. Generally larger than the right whale— it weighed more than 50 tons—the sperm could be as agile and wily as a salmon. The ponderous creature could come roaring up from a dive and leave the water almost completely. It could stand on its tail in the water, thrashing its powerful flukes with such force that its body could rise half out of the water and turn in a bobbing circle as it surveyed its surroundings.

Although the whales' eyes provided only lateral vision, sight was not the sense upon which the whales most relied: they dived so deep for their food that there was total blackness around them except for the faint light given off by many phosphorescing deep-dwelling creatures. Hearing, on the other hand, was vital to the whales, and the small size of their ears was deceptive. Sperm whales organized their whole lives around a symphony of sounds. Although they had no vocal chords, they emitted sonic signals—by moving air rapidly back and forth through passages in their heads—to give information to other whales nearby. They also produced a series of rhythmic crackling noises at a very high sonic level to explore the ocean's configurations and find the edible deepwater denizens lurking around them. They heard and were able to locate one another at distances of over three nautical miles.

So acute was the sperm whale's perception of danger that many a whaleman believed, incorrectly, that there was some mysterious connection between the whale and what appeared to be a slight oil slick usually found in its trail; the moment a boat touched the slick, these whalemen claimed, the sperm whale sank beneath the surface. Actually, what seemed to be a slick was simply the flattening of the waves by the whale's passage, as in a ship's wake—and the whale's disappearance was undoubtedly due to some noise made by the whalemen.

Once submerged, the sperm whale displayed one of its most remarkable attributes: its ability to dive deeper and stay down longer than any other whale. The whalemen worked out a rude rule of thumb: a sperm whale could stay submerged about as many minutes as its length in feet. A 60-footer, for example, could remain beneath the water about an hour—and when it surfaced, whalemen said, it required about 60 spouts to regain its breath.

Then it might take off across the water. A sperm whale's usual pace was about three knots—but pursuers would be astonished to see it accelerate to 10 to 12 knots and then keep this speed up for nearly an hour, leaving behind the swiftest whaleboat. Its strength seemed inexhaustible. Averaging about 80 miles a day, the sperm whale covered vast reaches of the ocean in migration patterns that became relatively predictable. "Forty-forties," some whalemen called the Atlantic sperm whales, because their range extended all the way from 40° S. lat. to 40° N. They covered those thousands of miles in large family groups, usually consisting of a bull followed by his harem of cows and their young; thus the sight of one sperm whale's spout often was followed by many more.

Yankee whalemen claimed that on a calm day they could often hear sperm whales before seeing them. The splash as one rose from the water

A mother sperm whale rears from the sea to cradle her mortally wounded calf in her jaws in this mid-19th Century watercolor. On occasion, whalemen would harpoon a baby whale, hoping to lure the fiercely protective mother into range.

and fell back on its side was thunderous, and the sounds of an entire herd of sperm whales spouting and lobtailing—slapping the water with their broad tails—reverberated across the sea for many miles. Sperm whales were frequently subject to colic and stomach ulcers, and their belly rumblings and belchings added to the cacophony.

A herd of sperm whales could contain anywhere from a dozen to a hundred animals. And theirs was a highly structured society. Solicitous as well as gregarious, female sperm whales would assist one another in danger, a pair sometimes supporting an ill or wounded companion to keep it from drowning.

The only time the males displayed aggression toward one another was at mating time each spring. When nature's time clock signaled the reproductive process, sperm whales fought titanic battles, and a whaleman who witnessed such a contest never forgot it. A young bull would challenge the sultan of the harem, as the whalemen referred to the leader of the herd. The cows would swim to the periphery, watching the spectacle. The two bulls would approach each other at full speed and meet head on with a cataclysmic impact. Turning and pounding each other with their tails, they would lash the ocean into foam. Attacking head on again, one would seize the other's jaw in his own, and the two mammoth bodies would roll over and over in a swirling sea of blood. Usually such an encounter would end in defeat for the young challenger, who would swim painfully back to the rear, gasping loudly for breath, as the herd reformed and went on its way.

But, inevitably, the older bull would one day be vanquished. He would then wander off from the herd and for the rest of his life swim alone. Perhaps 50 years old when driven from his herd, he could survive yet another 25 years, following similar lifelong migration patterns but in self-imposed isolation from all other sperm whales. The whalemen called these deposed monarchs old bachelors, and often found them easier to catch, as if the solitary creatures were at last weary of life's incessant struggle.

To the victor in such herd contests went the harem. Occasionally the whalemen also encountered the awesome intimacies of mating sperm whales, with a 50-ton monster playfully leaping out of the water to impress his partner, followed by an hour or more of splashing, rubbing and nuzzling before the brief but momentous climax as the two huge bodies rose from the water in a perpendicular embrace.

Rarer was the sight of a whale's birth, and even the most hardened of whalemen who came upon such an event were touched by the scene. The herd formed a reassuring circle about the mother, communicating with her and among themselves in shrill buzzing sounds, while she contorted her body and expelled the 12-foot baby, rolling over to snap off the umbilical cord. With the help of one or two other cows, she quickly nudged her newborn infant to the surface. Uniquely among ocean mammals, the baby whale was born tail first, undoubtedly because of its need to surface and breathe immediately after birth. Nourished by its mother's rich milk—which was 35 per cent fat—a baby sperm whale would nurse for almost two years, by which time it would weigh 40 tons and devour two tons of squid and fish each day.

Impressive as it was in the water, the sperm whale was even more astonishing when captured and dissected. At first whalemen thought that there were at least two different species of sperm whale, because the females averaged less than half the size of the males. For all its bulk, the sperm whale had an outer skin merely a couple of inches thick, usually serrated and scarred with old wounds inflicted by giant squid tentacles and alive with 16-inch sucker fish and colonies of stalked barnacles that the whalemen called dead man's fingers.

The yellowish mantle of blubber under this thin outer skin, though only a foot thick, was so strong that it could be cut and peeled from the heavy body without ripping or splitting. In addition to being a fatty reservoir for nourishment, it was a protective cylindrical layer of insulation. This blanket, as whalemen appropriately called it, was 75 per cent oil and covered the whale from head to tail, protecting it from underwater pressures and preserving its warm body temperature at 98.6° (the same as a human's). Beneath this thick sheath were more wonders: a gullet large enough to swallow a grown man (as Jonah was swallowed by the great fish of the Bible); three stomach compartments to digest the enormous chunks of rubbery squid (one sperm whale's stomach contained a 397-pound squid that was longer than a whaleboat); a quarter mile of intestines swarming with 20 different kinds of parasitic worms; a 300-pound heart pumping 15 gallons of blood 20 times per minute through an aorta a foot in diameter. "The aorta of a whale," remarked an 18th Century philosopher named William Paley in some amazement, "is larger in the bore than the main pipe of the water-works at London Bridge."

But the unique wonder of the sperm whale was its great lopsided battering-ram of a head. Only a quarter of the outside case was bone; the rest was sinewy flesh so hard that a harpoon would bounce off it. Inside this nearly impenetrable chamber lay the treasured reservoir of rich spermaceti—the first sight of which moved Nantucketer Peleg Folger to exclaim (according to an 18th Century volume): "His brains is all oyl!" A large whale might have 500 gallons of the oil, whose purpose was to provide a form of ballast. When the whale died, this oil solidified on contact with air into a waxy substance that could be pressed into the finest of candles, long-burning and virtually smokeless.

John Adams, while American minister to England in 1785, observed that "the Spermaceti oil gives the clearest and most beautiful flame of every substance known in nature," and suggested with New England practicality that it should find an excellent market in Europe.

All of this was not to mention ambergris, that much-prized rarity, of which large chunks were found in the bowels of some whales. Formed by an accretion of whale excrement around a squid's beak or some other indigestible matter lodged in the whale's intestine, ambergris emerged as a soft, black and evil-smelling mass. But on exposure to sun and air, it hardened, faded, began to exude a pleasant smell and exhibited astounding properties as a fixative for perfume. In the East it was used as a spice. So valuable was ambergris to the perfume and spice trade that in 1858 one whaling captain sold a single 800-pound specimen for $10,000.

Because of the sperm whale, what had been a minor enterprise along the New England coast for more than a century now rapidly became big business. For more than 80 years, sperm whale fishing remained dominated by Nantucket. The mainlanders for the most part continued whaling from the shore until by the mid-18th Century most of the supply of right whales had been depleted. The hamlet called Bedford Village, one day to become New Bedford, sent out only four long-distance sperm whaling sloops as late as 1765. Nor did the settlers on Martha's Vineyard, Nantucket's somewhat larger neighboring island, turn early to sperm whaling, though they would one day develop it into a major industry. They had the advantage of richer soil and were able to feed themselves from their fertile farms, and in the beginning they were content with that. But the Nantucketers, on their barren little sandhill in the ocean, had no such luck. For their bread, they had to farm the green pasture of the sea.

Swarming south and east into the open Atlantic, the Nantucketers found sperm whales in abundance. By 1740 some 50 sloops of 40 to 50 tons were bringing in almost 5,000 barrels of oil a year valued at $25,000. Eight years later Nantucket had a fleet of 60 sail, and its annual catch was valued at $96,000. In 1774 the Nantucket fleet of 150 sail was providing an income of $500,000 annually; each vessel was bringing in an average cargo of 150 barrels.

At first the Nantucketers cut the whale's blubber into chunks and brought it home in casks, as had the Basques, to be tried out in cauldrons on shore. By 1750, however, the Yankees had developed a tryworks that could be used aboard ship, and they were rendering the sperm whale's blubber into oil without coming home until their ships' holds were full of casks. This permitted them to go on longer voyages, lasting months at a time. Their little sloops gave way to much larger ships carrying whaleboats on davits. The double-ended whaleboat design was found to be the most seaworthy because following seas would separate harmlessly on the pointed stern, and the double end was also valuable when whalemen had to suddenly back their oars to move away from a threatening whale. The Nantucket whalemen found concentrations of sperm whales—whale grounds, they called them—off Bermuda, off the coast of Guinea, near the Grand Banks, in the West Indies, across the Gulf of Mexico and among the Azores.

In the process, the Nantucketers charted the powerful north- and eastward-flowing current of the Gulf Stream. When some London merchants once asked Benjamin Franklin why British vessels were taking so much longer to sail across the Atlantic than the colonists' ships, Franklin turned to his cousin, Nantucketer Timothy Folger, who obliged with a chart of the Gulf Stream. Franklin passed it along to the merchants, who ignored it—and continued to marvel at the apparent magic of the Yankee mariners.

Nantucket's whalemen not only dominated the actual capture of the whales but soon learned to control the commercial aspects of the industry as well. The merchant shippers of Boston had never attempted to compete with the Nantucketers in chasing the whale; there was quite enough money to be made in Boston as an entrepôt through which Nantucket oil, spermaceti and ambergris spread out to the world. Whale products went from Nantucket to Boston and from there to London, where the market was richer than in the still-struggling American colonies. The canny whaleship owners of Nantucket soon realized, however, that there were vastly greater profits to be realized from shipping directly to London, and also a good business in bringing back products from Europe that otherwise had to be purchased through Boston middlemen.

Indeed, Nantucket whalers were soon returning with products not only for the island, but also for Boston itself. One of these Nantucket vessels, the *Beaver*, sailed into Boston harbor on November 8, 1773, along with the *Dartmouth* out of New Bedford and the Boston merchant ship *Eleanor*, loaded with return cargoes of English tea. It was this tea, heavily taxed by British authorities, that angry American colonists dumped overboard in the famed Boston Tea Party.

The Nantucketers had their own reasons for disliking their British overlords. Throughout most of the century, England had been content to import sperm oil from the colonies. In 1770 there were fewer than 50 English whalers in the entire Atlantic, while Nantucket had 125. The price of sperm oil, meanwhile, had risen from £7 per ton in 1725 to £40 per ton in 1770. Whaling was grossing more than £150,000 a year on Nantucket, supporting nearly the entire population of 4,500 people. But as Britain confronted the growing possibility of alienation from her colonies, Parliament attempted to revive England's own whaling industry—both by offering bounties to British whalemen

Although he was a cousin of Benjamin Franklin's and a friend of John Hancock's, Timothy Folger, shown here as a well-to-do Nantucket merchant in a copy of a portrait by J. S. Copley, could not abide being a rebel against England. In 1785 he led 300 Loyalists, many of them whalemen, to Nova Scotia and set up a whale fishery—much to the delight of the Nova Scotians, who had been forced to pay British taxes on imported American oil.

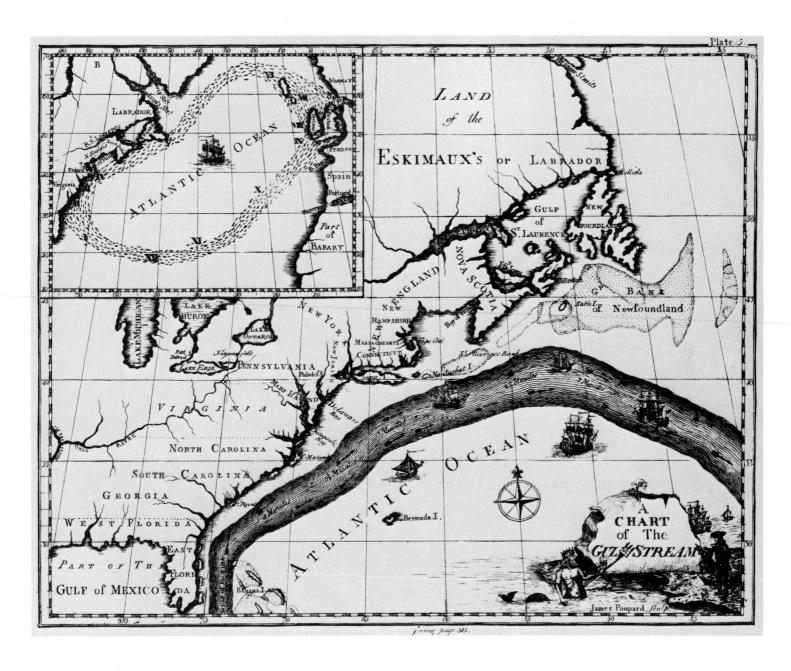

Drawing on information provided by his cousin Timothy Folger, Benjamin Franklin published this map of the Gulf Stream in 1786, superimposing its path on an old chart of the Atlantic Ocean. The inset shows patterns of herring migration.

and by hindering colonial whalemen with heavy duties on their oil.

Parliament also passed legislation attempting to limit the waters in which American colonists could hunt for whales, forbidding them the rich grounds off Newfoundland. But it was impossible, even for the British Navy, to police all the oceans. In April 1775, a Nantucket captain, Uriah Bunker, returned in the *Amazon* to report the discovery of fabulous new whaling grounds off Brazil.

That same month, the first shots were fired at Lexington and Concord. No greater calamity could have befallen the Nantucketers. The American Revolution devastated the island's whaling fleet. One by one, over the course of the next three years, nearly every whaler was laid up, or captured or sunk by British warships and privateers. Captured Yankee whalemen were given the choice of serving under the British flag or

An Old World haven for New World whalemen

In the late 1780s a debt-ridden London dilettante named Charles Greville devised an extraordinary scheme to build up England's whaling industry, as well as to recoup his own fortunes. He would build a town on his uncle's estate in western Wales—land he would one day inherit—and invite a colony of American whalers to come over.

Greville had little difficulty selling the plan to his uncle, William Hamilton, once he pointed out the benefits of his scheme. The estate, boasting a magnificent but undeveloped harbor, was a natural site for a port town. All that would be required, Greville argued, was a modest outlay for wharves, a customhouse and possibly an inn.

To members of Parliament Greville stressed the "national object" of the plan. The particular settlers he had in mind were Americans, mostly Nantucketers who in 1785 had taken up residence in Dartmouth, Nova Scotia, and were circumventing stiff British tariffs levied on imported American whale oil. By bringing these whalemen to Britain, Greville argued, the government would not only capture Yankee whaling expertise but also deal a blow to the tax-dodging colony at Nova Scotia.

Parliament apparently liked the imperial ring of the scheme. In 1790 the government approved establishment of the town of Milford Haven and in 1791 offered prospective settlers an inducement of £50 per family.

By 1793 Greville had attracted some 20 families to his port. The industrious newcomers set to work pegging out the lines for their homes, and soon they were sending their ships out into Atlantic and Pacific sperm whale grounds. Greville's building projects progressed more slowly, and his uncle became increasingly skeptical as the family resources dwindled. After the quay and a customhouse were completed, Hamilton flatly refused Greville's suggestion that he mortgage the estate to fund more projects.

Forced to look elsewhere for capital, Greville in 1796 talked the Navy into leasing land in Milford Haven for a shipyard. The town got another boost in 1802 when Admiral Horatio Nelson visited the Haven at Greville's invitation. The famous naval hero was treated to a cattle show and regatta, then obliged his host with a speech about the port's unequaled commercial potential—relying all the while on notes scribbled in Greville's unmistakable script.

Over the next few years, Milford Haven's streets continued to fill with houses, its harbor with whalers. And though Greville himself never got rich from the scheme, he was confident that his whaling town, at least, would prosper.

He was wrong. With Greville's death in 1809, the town lost its best advocate. Gradually, rival London whaling interests persuaded Parliament to withdraw the whalers' financial support. The War of 1812, meanwhile, wiped out nearly 20 per cent of the town's whaling fleet; two years later the Navy leveled another blow by moving its shipyard. Milford Haven took decades to recover—and then not as a whaling port, but as a local marine-supply and shipbuilding port. Some of the Americans adopted other trades; still others went home—not to Nova Scotia, but to Nantucket.

Belying the backdrop of prosperity, the near-empty harbor attests to the decline of Milford Haven's whaling industry in this 1812 scene.

risking imprisonment in prison ships; knowing the conditions aboard the latter, many whalemen chose the former.

As a consequence, although Britain lost the War of Independence, British whalemen gained a decisive advantage over their Yankee rivals, which Parliament attempted to make secure in 1784 by imposing a crushing £18 per ton duty on American whale oil. Because London was the commercial capital of the world, this duty had the effect of throttling American attempts to revive their industry. Moreover the British mounted a campaign to encourage Nantucketers to desert their homeland and join England's newly resurgent whaling industry. A prime mover was Charles Greville, who helped to conceive a scheme to establish a colony of Yankee whalemen, with their wives and children as well as their remaining ships and whaling gear, in Milford Haven, Wales *(page 56)*. With American whaling in the doldrums, many Nantucketers listened to the alluring British offer. Within half a dozen years following the end of the Revolutionary War, the British whaling fleet included more than 150 Yankee captains and 500 whalemen. The number of whalers flying the Union Jack soared from 50 in 1775 to nearly 200 in the early 1790s.

American-manned British ships extended their cruises across and down the Atlantic, reaping the profits of England's enforced monopoly. The growing demand for sperm oil sent prices to £95 per ton by 1786. The insatiable worldwide demand for sperm oil was so great that Britain's monopoly actually began to work against it. The British, even with their complement of Americans, could not fill the demand. Gradually, despite the duty, Nantucket began to get back into whaling: men who had given up whaling went back to sea, and they profited increasingly when the British lowered the tariff from £18 to £15 per ton. Moreover, numbers of transplanted Americans were becoming disillusioned with their lot as whalers for the British: their pay was less than expected and for a variety of reasons the colony at Milford Haven languished. The Americans returned home to join other Yankee whalemen in a rush to restore New England's whaling preeminence.

The great problem for both British and American whalemen was that the demand for oil was having a sudden and disastrous effect on the known Atlantic population of sperm whales. By the late 1780s, sperm whaling entered a decline, with fewer and fewer whales found on the once-teeming grounds between 40° N. and 40° S. Some captains came home after a year on the ocean with scarcely enough oil to pay expenses. Voyages grew longer and longer, as captains desperately drove their whalers to the far ends of the Atlantic in search of new grounds for the sperm whale. And one such cruise changed whaling even more surely than had Kit Hussey with his discovery off Nantucket of the mighty sperm whale a century earlier.

Poking her blunt nose south through the Brazil Grounds in late 1788, the British whaler *Emilia* found dishearteningly few whales. After a few weeks Captain James Shields ordered the *Emilia*'s course set far to the south, to the chill, barren Falkland Islands off the tip of South America. Still, lookouts in the crosstrees sighted little quarry. As the ship faced the grim prospect of returning home with a hold nearly empty, legend

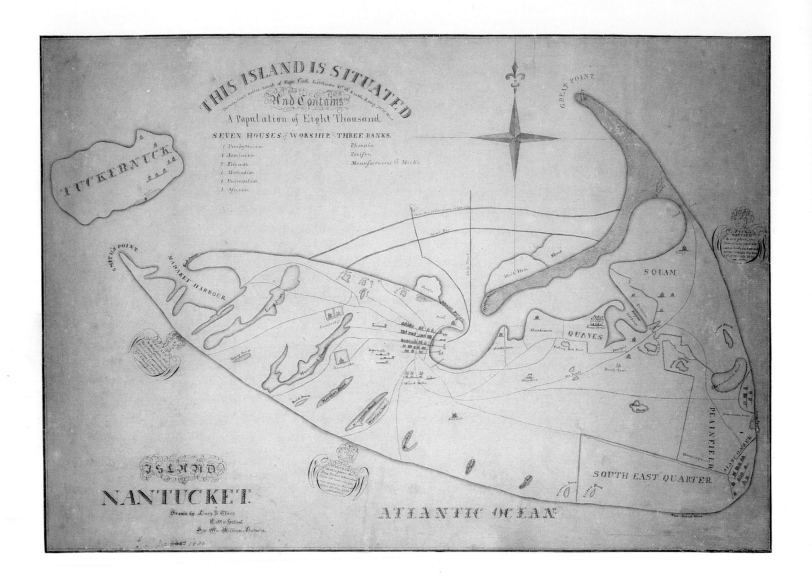

has it, the *Emilia*'s first mate, a Nantucketer named Archaelus Hammond, at last offered a bold suggestion: why should they not force a passage around Cape Horn and try their fortune in the Pacific Ocean? The whalemen had listened to merchant skippers describe great numbers of sperm whales in the Pacific, yet they had never acted on the information. For one thing Atlantic whaling had been — until recently — profitable enough. But now it was worth the gamble.

There is evidence that the suggestion did not come out of the blue. Shields probably intended to try the Pacific if his luck was bad in the Atlantic. Before the *Emilia*'s departure from London, her owners had written to the British Board of Trade that the ship had been outfitted "at a great expense to go around Cape Horn." The letter suggested that a number of other owners were eying the Pacific but that they had "declared that they shall wait till they hear whether our ship is likely to succeed there."

In any case, whether it was a spur-of-the-moment idea or a carefully thought-out plan, the voyage was a venture fraught with risk. No whal-

Thousands of years of action by wind and wave have shaped the tiny island of Nantucket until it literally points toward the open sea and the whaling grounds where most of its 19th Century inhabitants earned their living. The parallel lines across the harbor mouth in this 1830 map by a local schoolgirl named Lucy Macy indicate the irreversible formation of a sandbar that eventually cost the island its preeminence as a whaling port.

ing captain had ever tried to fight his way through the storms off Cape Horn, and the Pacific Ocean remained a mystery to most. In any event, the Horn passage proved as difficult as advertised. Although it was early summer in those latitudes, the *Emilia* had to battle her way through tremendous head seas and skirt frightful fields of ice before she gained the Pacific. Captain Shields reported that he was in "hard gales of wind for near 21 days." By March 1789, however, the *Emilia* was running up South America's west coast, all sails set, with dolphins wheeling around her in a sparkling sea.

Shields did not seek a harbor along the west coast of South America "without hazarding both the ship and lives, as it blowed a gale of Wind the whole time." The first place into which the *Emilia* sailed for a long-awaited landing was the Island of Massafuero, a few hundred miles off the coast of Chile. Here the men found fur seals "lying as thick upon the beach as they could be clear of each other," and many goats. They caught as many fish in two hours time "as lasted the crew 4 days."

And then it happened. While cruising off the coast of Chile there came the electric cry, "Bloooows!" And suddenly the *Emilia* was surrounded by sperm whales. It was a sight none of the whalemen had seen since before the War of Independence when the Brazil Grounds had first been discovered by Captain Uriah Bunker. Without a moment's loss, the whalemen tumbled into their whaleboats and rowed—first furiously, then stealthily—into the midst of the whales. And there on March 3, 1789, off a desolate coast, riding the swells of the chill Humboldt Current, Nantucketer Hammond achieved the distinction of being the first Yankee to harpoon a sperm whale in the Pacific.

For the next few months Captain Shields cruised up and down the west coast of South America, scarcely believing his luck. "I never saw so many large Sperma Coeti Whales all the time I have been in the business as I have this voyage," he wrote. At last, in September 1789, with every cask full of sperm oil, Shields took the *Emilia* thrashing around the Horn again. He put into Rio de Janeiro to treat 16 men who were ill with scurvy, and then set out for London.

The *Emilia* was running before the trades in the South Atlantic when she spoke the *Hope*, commanded by Nantucketer Thaddeus Swain. Captain Shields signaled to him to come alongside, so he could share the news of his discovery. But Captain Swain was in too much of a hurry; without shortening sail, he swept past the *Emilia* close enough so that he could bellow through his trumpet that he was 46 days out, bound for the South Atlantic grounds. Shields called back that he had come "from the Pacific by way of Cape Horn with 150 tons of sperm."

From the Pacific? One-hundred fifty tons? In a sudden change of heart, Captain Swain shouted to Captain Shields to heave to for a conference. But Shields was angered by his colleague's peremptory manner, and had ordered his ship held on course. The *Emilia* sailed on.

Captain Swain had heard enough, however. Without pausing in the South Atlantic, he took his *Hope* around the Horn. In a few months he was back in the Atlantic, headed for home with a full ship. This time, when he spoke the New Bedford whaler *Rebecca*, he paused for a short gam, as the whalemen called a visit between ships. He learned that the

Emilia's news had galvanized the entire industry. The *Rebecca* was bound for the Horn, and so were a swarm of other Yankee whalers. Captain Swain assured the *Rebecca*'s captain, Joseph Kersey, that there were plenty of sperm whales for everyone along the coasts of Peru and Chile, all the way from 30° S. lat. to 8° S.

By the time the *Rebecca* sailed for home several months later with more than a thousand barrels of oil, she had encountered no fewer than 39 other whaleships in the Pacific. The Nantucket ship *Beaver* returned with 1,100 barrels of sperm oil, for a profit of $20,000—a fortune in that day. The *Falkland*, the *Mary*, the *Penelope* and the *Harmony*, all sailing under the British flag with Nantucket captains, filled their casks faster than they ever had before.

By the turn of the century numerous whaleships from England and America were racing for Cape Horn and the beckoning, teeming sperm whale grounds of the Pacific. The golden age of whaling had arrived, and the cluster of New England towns from whence came the whalers began to profit beyond their grandest dreams.

At first, Nantucket was the preeminent port, its wharves piled high to collapsing from the weight of thousands upon thousands of oil casks— prudently covered with seaweed to protect the staves from drying out and shrinking in the sun. By the year 1820, no fewer than 72 whalers were sailing out from the island's harbor every year, and Nantucket was supplying the world with as many as 30,000 casks annually of precious sperm oil to lubricate the gears of the burgeoning Industrial Revolution, and to later light the lamps of great cities in Europe and America.

Nantucket's most prosperous period was described by a chronicler. "There were not many millionaires, though the whale business did make a few such, but many individual fortunes running into the hundreds of thousands were acquired, and that was great wealth for the time. It was no unusual thing for a captain to retire with a competency after a few successful voyages while still well under middle age; and shipowners who had several ships out at once often cleared enough to retire on in a very few years when matters went well with their ventures."

But Nantucket began to lose its predominance as the century progressed. A sandbar across the harbor mouth restricted the size of ships that could enter the harbor at a time when larger and larger whaleships were being built for the Pacific grounds. Some of Nantucket's whaling industry moved to the deeper harbor of Edgartown on nearby Martha's Vineyard, but the majority of whalers moved across Nantucket Sound and Buzzards Bay to the new capital of whaling—New Bedford. By the 1850s, four fifths of all American whaleships—nearly 400 bulldog-shaped vessels, amounting to more than half of the fleet of the entire world—sailed from this town of only 20,000 inhabitants.

In 1857 New Bedford ships brought back six million dollars' worth of whale oil and bone, a thirtyfold increase in half a century. The town shipped out New England's fourth largest tonnage of goods, exceeded only by major entrepôts like Boston. Virtually everyone's livelihood depended on whaling, from the candlemakers to the blacksmiths who forged the lances and the harpoons (which everyone called irons).

Thousands of miles of whale line—that strong, light rope that was essential to the whale hunt—were produced at the New Bedford Cordage Company, shown here in full operation in the 1860s. The rope was twisted in the long shed, called the ropewalk, telescoping from the main buildings. It was composed of 51 hemp yarns, each of which could withstand a pull of at least 112 pounds yet together measured but two inches in circumference.

Seated atop a cask of whale oil, a New Bedford businessman named Samuel Leonard conducts the affairs of his busy refinery and candle works in this 1855 painting. Behind Leonard at right, boxes of spermaceti candles slide down a chute from the window of the main building, while around the corner, men wrestle a cask onto a wagon drawn by three horses hitched in tandem—a necessary contrivance for navigating the lanes and alleys of the waterfront area.

Whale oil even served as currency, and many school teachers and ministers were paid in casks instead of cash. Not only did the people of New Bedford make a living from whaling; they made a superlative living, and their town, a village in comparison with New York, London, Boston or Paris, was one of the wealthiest per capita in the world. Ralph Waldo Emerson, who lectured there, remarked that the New Bedford merchants "hug an oil cask like a brother."

Evidence of the golden age was everywhere, from the bustling wharves along the waterfront to the mansions on the hill. The streets swarmed with a varied throng of Quakers and South Sea islanders, Africans and Azorians. New Englanders in flat hats and drab homespun no longer looked twice at tattooed, half-naked savages with filed teeth. But every small boy stared with envy at the young men flaunting in their lapels the badge of the harpooneer: a chock pin from a whaleboat. In the block-long ropewalks crews of men twisted strands of cordage into line for whaleboat tubs. On the cobblestone streets the passage of horse-drawn carriages with iron-rimmed wheels was deafening, mixing with the sounds of the clanging smithies, the clattering riggers and the cooperages echoing from the mallets knocking barrel staves together. The candlemaking factories exuded the heavy scent of spermaceti. There were warehouses aromatic with sandalwood and tea to be sold to the wealthy shipowners, and musky with the odor of ambergris waiting to be shipped to the perfumeries of London and Paris—or to the Middle East, where it was considered to be a powerful aphrodisiac. At virtually any time of day or night, a crowd might burst from a tavern on the waterfront to roll toward the docks, singing in celebration of a completed voyage or cursing at the prospect of a sailing day.

On the hill behind the waterfront, above it all, the mansions of shipowners looked down on the bustling confusion of the town. Quiet gentility pervaded the drawing rooms papered with silk from China and adorned with the finest furniture from England and Europe. From the windows of his paneled library a whaleship owner could survey the scene below with satisfaction. It represented the rich return on many investments, and not a few gambles, of money and men. There had been losses of both. A ship costing from $25,000 to $40,000 might have gone to the bottom off Cape Horn; many thousands might have been wasted on a cruise that unaccountably had turned up few whales. The Seamen's Bethel on Johnny Cake Hill was lined with memorial plaques bearing such inscriptions as: "carried overboard by the line, and drowned"; "fell from aloft, off Cape Horn"; "lost at sea with all her crew"; "lost overboard near the Isle of Desolation"; "killed by a sperm whale"; "towed out of sight by a whale."

For profitable though it was, whaling was a dangerous living. "The whale is harpooned to be sure," remarked one 19th Century observer, "but bethink you, how you would manage a powerful unbroken colt, with the mere appliance of a rope tied to the foot of his tail."

Yet the Yankees of New Bedford in the mid-1800s were prepared for all such risks and sacrifices. They well understood, as novelist-whaleman Herman Melville put it, that the wealth of New England was "harpooned and dragged up hither from the bottom of the sea."

Dock hands toss forkfuls of seaweed onto barrels of whale oil that have been off-loaded from a returning whaler, the Sappho, in this dockside photograph taken in New Bedford in 1870. The wet seaweed kept the barrels moist, preventing the staves from drying out and shrinking—and leaking their precious contents.

God Almighty of the quarter-deck

EDWARD S. DAVOLL, captain of five whaleships between 1848 and 1861, read a stern speech before every voyage. "Don't let yourselves be heard to grumble in any way," he admonished the crew, and to his officers he declared: "When you call a man see that he comes. When you send one see that he goes."

"This side of land I have my owners and God Almighty," declared an adage favored by Yankee whaling captains. "On the other side of land, I am God Almighty." By "land" the captain meant Cape Horn, separating the Atlantic from the Pacific, and the saying described the absolute authority he exercised on a voyage 10,000 miles from home.

A more incongruous deity could scarcely be found. Standing on the deck of a whaler, in a rumpled suit, a battered hat edging his brow, the typical whaling master eschewed the trappings of authority so prized by many of his merchant and naval counterparts. But spit and polish would have been out of place anyway in the working, blubber-and-bone atmosphere of a whaler.

The Old Man, as he was called, actually ranged in age from 25 to 70-odd years. Typically, he was a tough, resourceful New Englander in his mid-thirties who had signed aboard his first whaler in his early teens as a lowly cabin boy or foremasthand, and had gradually worked his way aft, "through the hawsehole," as sailors put it, to the officers' quarters. He had become a harpooneer and then a mate before winning a command of his own. And because he was thoroughly experienced in his perilous profession, he knew that discipline and teamwork were the only guarantees against failure and perhaps death. Captain Silas Alden of the bark *Bruce* roared to his crew at the start of a voyage in 1842: "I allow no fighting aboard this ship. Come aft to me when you have any quarrels, and *I'll* settle 'em. *I'll* do the quarreling for you, I will."

Moreover, captains of whalers led by example, taking charge of a boat whenever whales were sighted. A vigorous young captain was frequently the best harpooneer and the best lancer in the entire crew. Whalemen long remembered the exploit of Captain Silas West, who took out after a herd of 10 sperm whales in the Pacific and, incredibly, killed all 10 of them.

None of the captains pictured on these pages had the good fortune to duplicate West's record. But they were in the main such excellent leaders, hunters and navigators that the pragmatic businessmen ashore paid them the ultimate compliment of quoting insurance rates for American whalers at half the price charged for British vessels.

CALEB KEMPTON, a whaling master at the age of 26, had made four successful voyages to the South Atlantic by the time he was 34. Then, having built up a sizable nest egg, he retired to a farm at New Bedford and, until his death at age 82, invested in owners' shares of 26 whalers.

URIAH COFFIN, although born in
Nantucket, ventured to Europe to command
British and French whalers in the early
1800s. However, he eventually returned
home, and in 1820 commanded the
first whaleship from New Haven to cruise
the Pacific whaling grounds, coming
back with a profitable 2,000 barrels of oil.

HENRY PADDOCK, a highly respected
Nantucket shipmaster, met a tragic death
in the Pacific with the Catherine in
1833. As a crew member recalled the tale,
they were reprovisioning in Valparaiso,
Chile, when the normally agreeable
Paddock apparently got raging drunk,
and struck and killed a local man—
a deed for which the captain "was cruelly
executed by the Spanish authorities."

BENJAMIN RIDDELL went to sea as a boy of 15 and won command of the Equator at the age of 27 in 1831. Shortly before taking the vessel to sea, he married his childhood sweetheart, Lydia Coffin, but like so many captains, he was really wedded to whaling: he spent only 15 months ashore in the next 14 years.

ELIHU COFFIN, who was one of the first Nantucketers to venture into the Pacific, made five successful voyages on four different whalers between 1819 and 1845. His modest claim to fame was the discovery in July 1833 of a previously uncharted and bountiful atoll in the South Pacific that later, as Nassau Island, became a provisioning stop for whalers.

BENJAMIN CLOUGH saw one of his finest hours while a crew member on the Sharon in 1842, when he almost singlehandedly quelled a mutiny in which the captain was brutally murdered. The Sharon's appreciative owners immediately rewarded him with his own command.

AMOS HASKINS (above) was one of the few American Indians to become a whaling captain. A member of Massachusetts' Wampanoag tribe, Haskins went whaling in the 1830s, and eventually rose to be a master, earning a fortune sufficient to settle his family in a cozy New Bedford house.

JASON SEABURY, the master of the Monongahela, was either the perpetrator or the victim of a classic hoax in 1852. While he was at sea, several newspapers published a letter under his name describing the capture of a sea serpent that measured more than 100 feet, "was covered with blubber" and yielded oil "as clear as water." Before anyone could confront the captain, he was lost at sea.

CORNELIUS HOWLAND, shown at right with a model of one of his ships, became a captain at age 25 and was renowned for his consistent success in the Pacific between 1827 and 1844, when he retired. He rarely came home with fewer than 1,500 barrels of sperm oil, and in 1838 returned with a near-record 3,000 barrels after four years on the Magnolia.

Chapter 3

A Yankee ocean 10,000 miles from home

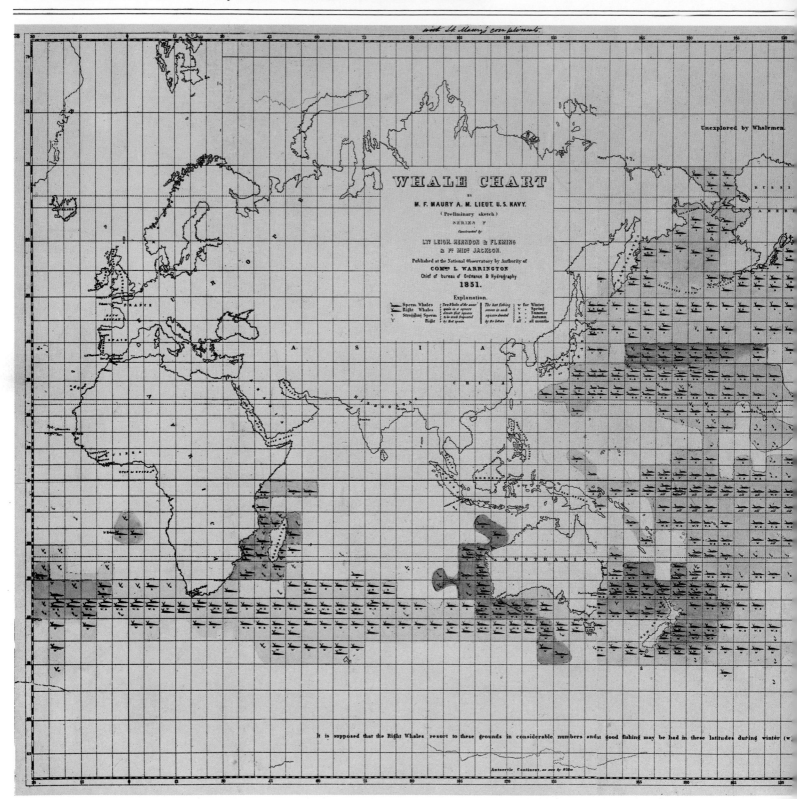

The Pacific teems with sperm and right whales in this 1851 chart by U.S. Navy oceanographer Matthew Maury. The Navy was seeking information on winds and currents, but to obtain whaling logs for his project, Maury had to promise whalemen in return a compilation of all their sightings. Maury's codes are explained in the key—except for the four mysterious blue areas.

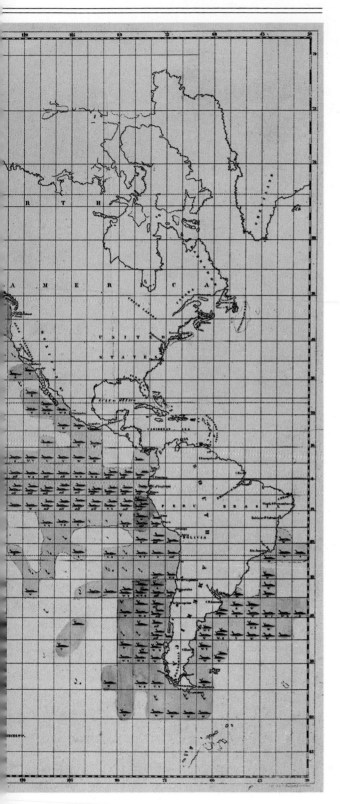

On Sunday, December 22, 1850, John Scott DeBlois, master of the whaler *Ann Alexander,* out of New Bedford for the Pacific hunting grounds, wrote a letter to his wife: "I got a whale that made 80 bbls. after great del trouble. He was one of the knowing kind. I struck him and the whale went off a little way and then came for the boat. He stove the boat ver bad so he seam to be contented with that and lade clost to me. The other two boats was off about 4 miles from me in chase of other whales. Finly the mate came to me. I told him never mind me but get the whale if he cold. He went and struck him and the whale stove his boat worst than mine, and heart 3 of the men. He nock the mate as high as our house.

"Well I begin to think that fish was not for me. He war to stoven boats and 12 men swimming in that sea and the Ship and other boat a long way off. Finly I got the Ship down to me and got another boat out and tride the whale agan. He tride his old triks agan but he cold not com it. And I got him.

"Now," concluded DeBlois, "you see what perseverance dos."

Perseverance was assuredly a hallmark of the Yankee whaleman, and nowhere was it more evident than in the vast Pacific during the golden age of 19th Century whaling. By mid-century the whale hunter averaged a grinding 42 months on each cruise on the great ocean. "There is his home," wrote Herman Melville. And in that faraway home, thousands of miles from his snug New England harbor, the whaleman's life was shaped by tedium and terror, by loneliness and exhilaration, by the pleasures of exotic islands and a head-on clash of cultures that all too often ended in violence. Only men of patience and fortitude could survive such voyages and muster the will to return again and again.

Some Yankee whaleships, to be sure, stayed in the Atlantic, questing for declining numbers of right as well as sperm whales on what Pacific veterans contemptuously called "plum-pudding cruises." Right whales were also found and hunted in many parts of the Pacific, especially on the northern and southern peripheries. In the early 1800s there were so many right whales off southern Australia and New Zealand that Tasmanian authorities warned boatmen in the Derwent estuary to "keep close along the shore, it being dangerous to venture into mid-channel."

But the mighty sperm whale was the real prize for the Pacific whalemen. Though they could not know it, possibly a million or more of the beasts inhabited the ocean. Yet so vast was its expanse that even those abundant numbers would have remained largely hidden had not the whalemen learned just where to hunt. The first captains around Cape Horn found concentrations of sperm whales all along the west coast of South America; those who came after pushed west to discover feeding areas throughout the Pacific and to chart the seasonal migrations of the whales to these favored parts of their domain.

As early as 1818 Captain George Washington Gardner in the *Globe* out of Nantucket had sailed west from the volcanic Galápagos Islands, which lie astride the line, and found a broad feeding area ranging from long. 105° to 125° W. between lat. 5° and 10° S. Oil from these Offshore Grounds, as they came to be known, filled the casks of countless Yankee whaleships over the next 50 years.

At about the same time, an American merchant captain spotted numerous sperm whales farther to the northwest in a patch of ocean several hundred miles southeast of Japan. He tipped off some Nantucket friends and set off a rush to those feeding grounds. Among the first to arrive in 1820 was Captain Joseph Allen in the *Maro*, which laid down 2,425 barrels of sperm oil in the course of 30 months. Within three years, more than 60 whalers were prowling the Japan Grounds. In rapid succession the Yankee whaling masters found other sperm whale haunts off the Society, Samoan, Kingsmill and Fiji Islands, all within the broad equatorial belt. They also fished elsewhere in the South Pacific, between 21° and 27° S., and in the North Pacific from 27° to 35° N.

A persevering captain—to use Captain DeBlois's word—would cover many of these grounds in the course of a single cruise. The pattern varied depending on the time of year and on reports, rumors and each master's whim. But a typical voyage after the 1820s would bring a whaler around the Horn into the Pacific in October or November to cruise among the Galápagos and perhaps west to the Offshore Grounds. Then she would be taken into Honolulu or Lahaina in the Sandwich Islands for provisioning and repairs before setting out for the Japan Grounds. Returning along the equator, the captain had an infinite number of possible routes north or south of the line. The pattern could be altered or repeated for years, with an occasional return to the Sandwich Islands to unload and ship home the whaler's oil before setting out again.

In their wanderings the whaling captains came across dozens of obscure islands and atolls: such Yankee-named outcroppings as New Nantucket, Gardner, Swain, Chase, Rotch, Coffin, Starbuck and Independence Islands appeared on the charts that whalemen exchanged with one another. Indeed, 19th Century European cartographers began designating the entire central Pacific as American Polynesia in recognition of the dominant Yankee presence.

Captain James Shields in the British whaler *Emilia*, of course, had led the way into the Pacific in 1789. But during the War of 1812, the U.S. frigate *Essex* had rounded Cape Horn and captured a dozen British whaleships, a setback from which British whaling in the Pacific never fully recovered. By the 1820s, 120 American whaleships were working the Pacific. In 1846, the peak year, the Yankee whaling fleet numbered 736 vessels valued at more than $21 million, and at any given time most of this investment was to be found concentrated in the Pacific.

The rewards were not just handsome—they bordered on the stupendous, with whalers consistently bringing home profits from a single trip that doubled or tripled the money their owners had invested in the ships. The *Uncas* of Falmouth returned from the Pacific with 3,468 barrels of sperm oil worth $88,000; the *Sarah* of Nantucket came home with 3,497 barrels worth $89,000; the *Magnolia* of New Bedford brought 3,451 barrels worth $85,000—all on an average investment of less than $30,000 for a fully rigged and equipped whaleship. Since two thirds of the oil went to the owners, it was not long before those worthies got rich.

Yet in terms of human life and hardship, whaling exacted a fierce price for the profits of the halcyon Pacific years—as witness the cruise of the Nantucket whaler *Franklin*.

Nantucketer Frederick Coffin points with a discoverer's pride to a chart of the prolific Japan sperm-whaling grounds he helped pioneer in 1820. His ship, the Syren (right, foreground), sailed to her home in England in 1822 with 2,768 barrels of oil, a haul that, as one contemporary noted, "astonished and stimulated to exertion all those engaged in the trade throughout Europe and America."

On June 27, 1831, the *Franklin*, with about 25 men aboard under Captain George Price, left for the Pacific and—as later related in fragments by Nantucket historian Obed Macy—a journey into melancholy.

Soon after the *Franklin* left home, a crewman was hurled from aloft during a storm, suffering severe injuries that laid him up for two months. On November 15 another seaman tumbled from the mizzen-topgallant head, breaking both legs. Sometime later, a sailor suffering from consumption—which flourished in the dank, cramped, airless forecastles of whaleships—was put ashore at Callao, Peru, where he soon died. In mid-February, 1832, a line got tangled as one of the *Franklin*'s boats put a harpoon into a whale; when the whale sounded it carried the boat down, drowning two men.

A year went by without further recorded mishap. However, in February 1833 yet another seaman died—in a fall from aloft, where he had been perched searching the horizon for whales. Six months later, in August, a harpooneer was caught up and carried away by a line he had made fast to a whale.

The next month the *Franklin*'s log showed its most ominous entry to date: one of the hands died of scurvy, a foul disease to which whalemen were especially vulnerable because of their prolonged stays at sea. Among the symptoms were putrid breath and spongy, ulcerated gums, followed by painful swelling of the joints and—finally—massive hemorrhaging. The malady, caused by vitamin deficiencies, could be swiftly remedied by eating fruit, and when a whaler touched one of the islands the men gorged on fruit. But such interludes were infrequent because island visits interfered with the business of hunting whales, and the monotony of salted meat, hardtack, beans and rice was usually broken only by such dubious delicacies as fritters of whale brains mixed with flour and fried in whale oil, and barnacles, which were said to be especially tasty when taken from the nose of a whale. Neither of these, alas, was of any help against scurvy.

Eight months passed, and then on June 3, 1834, the ship's mate died after he had, according to the ship's log, strained himself getting terrapins. The exact nature of the accident was unreported, but it doubtless occurred while the unfortunate mate was trying to manhandle one of the giant Galápagos tortoises, commonly miscalled terrapins by the whalemen. Taken alive, the tortoises suffered scarcely any loss of weight even when unfed for weeks, and they provided a welcome addition to the whalemen's drab diet.

On June 8, 1834, a bad day even for the *Franklin,* both Captain Price and the ship's steward died. The causes of death were not immediately recorded. But over the next few days three more men died—and the log recorded that all five had perished from scurvy. A sixth man died of scurvy a month later. That brought the total to 13 dead and two injured.

The next death was that of the whaleship itself. On October 5, 1834, while on her way home in heavy weather, the *Franklin* fetched up on a reef off the coast of Brazil. All hands survived, but the *Franklin* and two thirds of her tragically hard-earned cargo of whale oil were lost.

Practically all of the *Franklin*'s misfortunes were commonplace on whalers—and they by no means exhausted the ways in which Jonah's luck could come to a Pacific whaleship and its crew. Just to get to the Pacific grounds was difficult and dangerous enough. Even though some cautious captains traveled the longer and slower but relatively safe route around Africa's Cape of Good Hope, most captains were willing to take a chance on Cape Horn, at the tip of South America, in order to gain more time on the whaling grounds. This meant sailing the bluff-bowed whaleship against head seas that had swept 4,000 miles across the Pacific and now came thundering eastward through the comparatively narrow channel between the Patagonian Cape and the ice shelves of Antarctica. Winds that howled down from the Andes formed tight cyclonic patterns in which a favorable slant could shift to a head-on storm in the space of a few minutes.

Even in summer the weather was sometimes so cold that blankets froze in the forecastle and ice crackled from the rigging. Winds reached 100 miles per hour and waves rose 50 feet, often burying the deck under tons of green water. The whalemen, their hands swollen with chilblains and red with frozen blood, had to face away from the gusts in order to breathe, and the whaler reeled until her yardarms were close to dipping in the waves.

Numbers of whalers were lost at the Horn. And more than a few exper-

Two memorials reflect the hazards that beset Yankee whalemen in far-off seas. The elaborate painting of mourners standing next to Ebenezer Gifford's cenotaph was probably commissioned by a family member; the simple plaque for William Kirkwood was ordered by his shipmates and mounted on the walls of the Seamen's Bethel in New Bedford.

ienced a battering similar to that of the Nantucket whaler *Harvest*, which attempted to round the Horn on November 14, 1844. She had been battling a storm for days when a mountainous sea overcame her. The gigantic wave crashed over her bow, carrying away her mizzenmast, all her boats and all the hapless men on deck watch—the second mate and eight hands. The *Harvest*'s demoralized captain put back into the Atlantic and never again, so far as anyone knows, attempted the Horn.

After Cape Horn the Pacific Ocean, with its warm trade winds, phosphorescent waters and playful dolphins, seemed like paradise. Yet here in the months of August, September and October the placid waters gave birth to terrible typhoons, worse than any Atlantic hurricanes, with winds of 150 to 200 miles an hour. And there were other dangers all the more insidious because they were hidden. Until the coming of the Yankee whalemen, the Pacific was still largely a *mare incognitum*, at least in a hydrographic sense. Dozens of whaleships were lost on uncharted reefs. The *Canton* broke her back on a coral head in the central Pacific in 1854, and her crew drifted in their boats, with half a pint of water and half a biscuit per man per day, for 90 days, until they reached Guam some 3,000 miles away.

Sometimes the danger was even more disguised. The merest brush of a ship's bottom against an unseen coral reef could peel away copper sheathing already weakened by the action of salt water. The naked planking was then open to onslaught by countless whitish, wormlike mollusks, properly named teredos but popularly called shipworms. Capable of attacking even the hardest wood (teak was the only exception), the teredos entered the ship's planks while young, making pinhead-sized holes; once inside, the animals enlarged their burrows as they themselves grew—to lengths ranging from a few inches to three feet. Thus planks that to outward appearances seemed sound might in fact be so honeycombed that the wood could be crushed by a man's fist.

Such was the fate that befell the *Minerva* of New Bedford. In August 1856, off the Kingsmill islands in the Gilbert Islands, the ship scraped a coral reef so gently that Captain Calvin Swain scarcely noticed it. He was mystified when, six months later, the ship began to leak at a rate that steadily increased to 250 pump strokes per hour. Captain Swain made for Norfolk Island, but contrary gales drove him south toward Sydney, Australia, while the pumps went to 2,400 strokes with 16 inches of water pouring in each hour. Clearing out the forehold and standing in water to their waists, the men managed to stanch some of the flow with tarred canvas and blankets, and the *Minerva* struggled into Sydney. There, when the ship was hauled out, Captain Swain found that the copper sheathing beneath the water line on the starboard bow had been scraped free and that shipworms had reduced the planking to a shell. It took expensive and time-consuming repairs before she was fit to sail again.

On board the whaleships, even routine chores carried hazards peculiar to the business. No matter how diligently the men swabbed, or swept with brooms made of the bristles of certain Pacific island plants, decks and companionways often stayed slick with oil and gurry, causing many a painful or disabling fall. Fire was a constant threat whenever a whale was being tried out. With blubber-fed flames sputtering and blazing beneath the try-pots, it did not take much to touch off a conflagration. A sharp roll of the ship could slop boiling oil onto the fires with explosive effect. Even a spark could set a spreading fire on the oil-soaked decks.

Of less immediate threat to life and limb, but almost as hard to live with, were the vermin that—attracted by whale oil and blood—flourished on whalers as on no other vessels. A powerful and dedicated ship's cat might keep the rat population within reasonable bounds. But cockroaches were quite another matter. The frigid weather off the Horn appeared to kill some of them off, or at least render them dormant. However, in the warm Pacific they reappeared in scuttling armies, with individuals growing up to an inch and a half long. They had an especially offensive predilection for nibbling around the lips of sleeping men, presumably in quest of food residues.

In their reminiscences Pacific whalemen seemed to take a perverse pride in their formidable cockroaches. Crew member Francis Olmsted boasted that the insects so infested his vessel that he had actually seen one "up in the main top, wandering about very much at his leisure." In their mass peregrinations about the ship, Olmsted said, the cockroaches made "a noise like a flush of quails among the dry leaves of the forest. They are extremely voracious, and destroy almost everything they can find: their teeth are so sharp, the sailors say, that they will eat off the edge of a razor."

All the problems of shipboard life were aggravated for whalemen because of the long periods they spent at sea. Unlike merchant and naval vessels, which always seemed desperately anxious to get somewhere, the whaleships were in no hurry. Once a whaler had reached a hunting area at what its captain reckoned to be the right time of year, there was little to be gained by tacking busily about the ocean; it was easier, more economical and just as productive to poke along under light sail until whales appeared within the 8- to 12-mile sweep of the lookouts' gaze.

Moreover, it was usually pointless to sail in darkness. (Exceptions to this were observed when a ship was traveling between whaling grounds, or when it was trailing a wounded whale or a healthy one that had been spotted at sundown: the animals, when not disturbed, moved at a fairly predictable rate and generally continued throughout the night in the same direction they were last seen taking.) Thus, almost alone among the deep-sea vessels of the world, the whaleships usually furled their topsails, backed the mainsails and lay hove to overnight.

At sunrise the sails were reset and the lookouts climbed to their precarious perches in the crosstrees 100 feet or more above the rolling sea, steadying themselves by using as handholds the horizontal hoops that enclosed them. Watching for whales was wearisome work, demanding intense concentration to overcome the blinding reflections of the equatorial sun thrown up by the water, and the entire foremast crew shared the duty, dividing it into two-hour stints.

A sperm whale normally sent up a single 10- to 12-foot spray of condensed moisture that hung in the air for about three seconds. It could be seen for about six miles. At that distance it was easy to make a mistake and, rather than set off a needless burst of activity on deck, cautious

With menacing cliffs close by her starboard bow, the New Bedford whaler Kutusoff fights through Cape Horn's turbulent seas in this scene from a panorama depicting the whaler's 1841-1845 voyage around the globe (pages 100-107). The passage from the Atlantic to the Pacific could be so difficult that there was only slight exaggeration in the wry comment of a captain's wife on another whaler: "We made 20 knots today — 10 straight ahead, and 10 up and down."

lookouts usually watched for a second spout before singing out the famous "Thar she blows!" or perhaps "There blows!" If the whale then leaped or thrust its head high above the surface, the lookout called, "She breaches!" And when the breaching whale fell back into the sea, he shouted, "She white-waters!"

But in the enormity of the Pacific, days could pass without a call from the crosstrees. Time weighed more oppressively on the expectant crew of a whaleship than on other sailors. On merchantmen seeking maximum speed, the sails were in need of frequent adjustment. But on a whaler the men could only wait—and hope. The harpooneers incessantly checked their gear, sometimes spending an entire morning sharpening harpoon and lance points that only the previous day had been made as sharp as a whetstone could hone them, or re-coiling lines that already lay perfectly within their tubs. For the other hands, once the routine shipkeeping chores were completed, there was little to do. Those who could might read; Francis Olmsted reported that his whaleship, the *North America,* had a library of some 200 books. Others devoted the empty hours to washing and mending clothes worn threadbare by long, hard use; whalemen referred to their pants and shirts as "a patch upon patch, and a patch over all."

A few fished, as much for amusement as for food; they used hooks baited with white rags, which skipped along the water's surface in simulation of flying fish, the favorite food of bonito and albacore. Yet of all the pastimes, perhaps the most popular was that extraordinary form of folk art practiced exclusively by whaling men: the carving of whalebone and teeth into scrimshaw. Talented scrimshanders, as they called themselves, could produce some astonishing objects (pages 126-131).

As evening fell upon the Pacific, the foremasthands usually grouped around the fo'c's'le scuttle in a community of spirit all the more remarkable because of their diverse origins. Their one common characteristic was youth. When the Hector sailed from New Bedford in 1832, 12 of her crew of 27 were in their teens; only five were over 25, and the oldest was 32. Otherwise, a whaler's crew presented a kaleidoscopic picture. New Englanders predominated in the early years. As the 19th Century progressed and an increasing number of Yankee lads answered the beck of the American West, many whaling masters took to detouring by way of the Azores on their outward voyages in order to fill out their crews with Portuguese islanders. These "Portagees," as they were called, made good hands, and there were always plenty of them willing to exchange conditions at home (which included compulsory military service) for life on board a whaler. So many Portuguese eventually settled in New England that by the end of the century one section of New Bedford was called Fayal, after one of the Azores, and there were more Perrys (originally Pirez) in the city than there were Smiths.

Another favorite recruit was the Sandwich Islander, whom the Yankees called a Kanaka, as they did all the other Pacific islanders. The Sandwich Islander, or Hawaiian, was an accomplished boatman, a superb swimmer, brave yet docile; not least important, what seemed like slave wages to a New Englander was wealth to him. By 1844 there were more than 500 Pacific islanders serving on Yankee whaleships, many as harpooneers. The Sandwich Island government licensed its recruits, and the captains posted a $200 bond, promising to return each man within three years. But the men were usually quite happy with the whaleman's lot, and most captains forfeited the bond, regarding it as a bargain price for acquiring a top hand. Since the captains had no desire to cope with their tongue-twisting names, the islanders were generally entered into the ships' logs merely as Joe, Jack, Jim or Sam Kanaka.

To these stalwarts were added adventurers and drifters from ports the world over—Spaniards, Swedes, Germans, Irishmen, Frenchmen, Italians. Many were landsmen who for one reason or another chose to sign on for a whaling cruise. Historian Elmo Hohman described the group in one whaler's fo'c's'le. Among them were a failed Shakespearean actor, a kleptomaniac Englishman, a homesick farm boy, a Philadelphia hardware clerk, a bullying stonecutter and a well-educated Kentuckian who had given up a job as a newspaper reporter covering the U.S. Senate to "see the world."

A mild Pacific night, with the sails furled or backed and the ship rolling gently to the swells, could help such men forget not only their

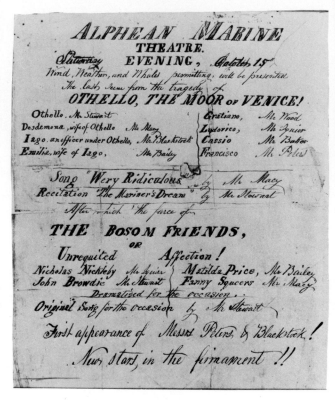

A playbill for an evening of theater on the whaleship Alpha in 1848, dreamed up by William Hussey Macy, the ship's cooper, advertises dramas, recitations and songs performed by the crew to enliven the long cruise. In a letter to his cousin in Nantucket, young Macy recounted how the men devised a stage on deck "with an immense drop curtain triced to the rigging. 'Tis astonishing what a display can be made of with such limited materials."

past but also the difficulties and dangers of their present. The shrouds hummed softly. The sea washed alongside. The mast tops reached toward the stars. In the chicken coop under the workbench aft of the tryworks, clucking hens settled down to lay eggs for the officers' breakfast. Pipe bowls glowed in the dim light. Tobacco was the whaleman's narcotic; it helped relieve the boredom of the endless search, and the average hand smoked or chewed nearly 100 pounds of the stuff a year. (More expensive cigars seemed to be the captain's prerogative, though they were not always smoked. Harry Chippendale, who later became a naval captain, recalled his bemusement as a boy, watching Captain George W. J. Moulton of the *Sunbeam* slowly chew his cigar down to a short stub without ever lighting it.)

Many men passed the hours telling stories, some true, some apocryphal or outright fiction. There was the yarn—probably true—about "Dukey" Robinson, who always stuffed a red bandanna in his mouth when chasing a whale: Dukey could not swim and he believed that keeping water out of his mouth would help him float if he fell overboard. Once a whale did knock Dukey overboard; he pulled himself along the harpoon line, clambered up on the whale's back and was found, without his bandanna but singing lustily, by the men who came to his rescue.

Another favorite story recounted how Captain Peter Paddock struck and lost a whale in 1802; 13 years later, in the same waters, Paddock killed a whale and found his lost harpoon, rusted but still embedded in the whale's hide. Then there was the story of the *Milton's* whale, which surprised everyone by whistling when it spouted. The explanation emerged later: while cutting into the whale's head, a man gashed his foot on a harpoon blade from the whaler *Central America* that had been stuck

Crewmen on a whaler scramble madly for their shares of salt junk, or pork, left in a wooden bucket on the fo'c's'le deck by the ship's cook. The 1846 engraving is only slightly facetious; though the food aboard a whaler was the "acme of indigestibility," as one observer put it, the cook's cry of " 'Meat! Meat! Fall to, all hands!' never required repetition." The men reacted "like hungry wolves," and the ship's boys slithered "like eels under the legs of those who were ahead of them."

in the whale's spout hole for 15 years. There was also the right whale captured in the North Pacific around 1870 by the *Cornelius Howland*. The men found an iron from the *Ansel Gibbs,* which for the previous 10 years had been whaling only in the North Atlantic. Right whales were never known to have crossed the equator, so this one could not have gone down around the Horn. Could it by chance have found the Northwest Passage? Many whalemen thought so.

The favorite legends by far were of whales that became famous for the fights they gave whalers. New Zealand Tom, named for his favorite feeding grounds, was a black sperm whale with a white hump on his back that made him recognizable; he destroyed dozens of whaleboats in the early 1800s. He was finally taken by a whaleship named the *Adonis,* but not before he had smashed nine pursuing whaleboats. And when he was at last slain and brought to the ship for flensing, the harpoons of several other whalers were found rusting in his hide.

Another legendary opponent was Timor Jack, who haunted the Timor Sea and evaded capture for many years, smashing countless whaleboats. He was finally taken, so went the tale, by an ingenious ruse: whalemen lashed a barrel to a line, towed it behind a boat and lanced Timor Jack to death as he furiously attacked the bouncing barrel.

But the most awesome whale of all was Mocha Dick. Although he passed into popular mythology, Mocha Dick was a real-life, flesh, blood and blubber bull sperm whale, named not for his color—more gray than brown, with a white, eight-foot scar across his immense head—but because his first reported attack was against a whaleboat near Mocha Island, off the coast of Chile, around 1810. Thirty years later Mocha Dick was still at it, his reputation by then so embellished through countless retellings of his story that even his color had become, in the words of one narrator, "white as wool." No one knows how many of the depredations for which he was blamed were actually committed by Mocha Dick.

The crowning series of exploits attributed to Mocha Dick began in the summer of 1840. In July the English whaling brig *Desmond* was working 215 miles off Valparaiso when a lone whale—the biggest the crew had ever seen—breached nearly his full 70-foot length about two miles away. When a pair of boats was lowered and started toward the whale it immediately rushed them, crashing head on into the lead boat, then seizing it in his huge jaws and reducing it to kindling. While the whale sounded, the second boat picked up the survivors—whereupon the monster came roaring up from the bottom, bashing into the boat and sending its occupants flying. When the *Desmond* came to the rescue, two men were missing. The others told of a large white scar on the beast's brow.

One month later and 500 miles to the south, two boats from the Russian bark *Serepta* killed a solitary whale and began towing it back to the ship. Mocha Dick breached spectacularly between the boats and the ship, then took dead aim on the boats. He smashed the first one and started for the second, but the mate maneuvered his boat behind the carcass of the dead whale. The *Serepta* picked up all hands, and as she sailed away Mocha Dick was still lingering in the vicinity of the dead whale—as if on guard.

How frustrating—not to say enraging—an unsuccessful voyage could be is evident from this blistering broadside issued by the crew of the Lancaster after they were skunked, as they put it, on a cruise in the 1850s. The men railed against the "ignorance, inefficiency and utter incompetency of the mates" and reserved their greatest scorn for Third Officer Bushnell, who lost a ripsack—a gray whale—after a six-hour struggle. About the first mate they say nothing save that they leave him to the mercies of the captain, whom they apparently consider blameless.

MASTERS
OF
VESSELS!

And all others interested, are hereby publicly cautioned against shipping the following officers of the "LANCASTER," of New Bedford, as it was through their ignorance, inefficiency and utter incompetency that the "Lancaster" was "SKUNKED!"

WILLIAM HENRY ROYCE,
SECOND OFFICER,

Was 3d mate and boatsteerer of the Bark "Black Eagle" for the season of '55, during which time he distinguished himself as an excellent DO-NOTHING, whilst as 2d officer of the "Lancaster" he won for himself the reputation of an extensive KNOW-NOTHING! Too ignorant to catch a bow-head, and afraid as death of a right whale. Would make a good deck walloper.

CHAS. BUSHNELL,
THIRD OFFICER,

Is equally incompetent and worthless. Was boatsteerer in the "Washington" when lost--no oil! Then 4th mate of the "Wm. Badger--no oil! Again 4th mate of the "Huntsville," brought no oil to the ship! And finally 3d dickey of the "Lancaster"--SKUNKED! Was fast six hours to a ripsack which drove him out of the head of the boat and from which he finally cut. Would make a good blubber room hand.

Of the mate we will say nothing, preferring to consign him to the tender mercies of Captain Carver.

Before shipping any of the above worthies, Masters of Vessels are requested to ascertain their true characters.

(Signed by the entire Crew.)

The next May the British whaleship *John Day* out of Bristol had made a kill east of the Falklands, and smoke was billowing from its try-pots when Mocha Dick breached scarcely 100 yards away. Three boats were put down. Dick swam away to windward, then turned, paused for a moment—and made his rush. He missed and, as he surged past, the mate in one of the boats planted a harpoon. Dick sounded, surfaced, took the boat on a wild three-mile ride and then turned to attack. As the mate desperately shouted "Stern all!" Mocha Dick struck the boat broadside, his huge body carrying completely over the craft; he flogged it to bits with his flukes and killed two men. Dick lay off a short distance and waited for the other two boats to come up. A man in one of the boats caught the trailing harpoon line. That seemed to enrage the whale and he attacked again, battering out the boat's bottom and flinging the men into the sea. Remarkably, no one was hurt, and the whalemen made their escape in the third and last boat.

More than a year passed without word of Mocha Dick. But then one

A mission of mercy to forbidden isles

The whaler Manhattan rides a choppy sea in this Japanese watercolor depicting her visit to Uraga Bay. The characters record the vessel's dimensions.

day in October 1842, he fought the most glorious battle of his long, combative life. Mocha Dick began the day with an apparently unprovoked attack against a schooner carrying lumber off the coast of Japan. He crushed the little ship's stern, but its cargo kept it afloat as the whale swam slowly out of sight to port. Shortly thereafter, three whaleships—the *Yankee* out of New Bedford, the British *Dudley* and the *Crieff* of Glasgow—cruising in company, came upon the damaged schooner and learned of the whale's onslaught.

Even as the whaling masters discussed plans for seeking and slaying the beast, Mocha Dick reappeared, breaching high a mile to windward. Each whaler lowered two boats—six in all. Dick sounded for 20 minutes. When he surfaced one of the *Yankee* boats planted a harpoon in the whale's body just behind the head. Mocha Dick let go a great, gasping spout; his flukes moved weakly and then he lay still, as if dead. The whalemen waited five minutes before warily approaching the inert body—whereupon Mocha Dick, who had evidently been acting, came to

U.S. Navy Commodore Matthew Perry is credited with being the first American to make serious contact with the Japanese, and to commence the process by which these isolated people were brought into the modern world. What is not generally known is that Perry had forerunners. A crew of Yankee whalemen reached Japan in March 1845, eight years before Commodore Perry, and they made a profound impression on the Japanese.

Captain Mercator Cooper and the 28 men of his Sag Harbor whaler *Manhattan* had shaped their course for Japan, seeking neither whales nor manifest destiny. They were, instead, on an errand of mercy. In the North Pacific they had rescued 21 shipwrecked Japanese sailors and had tried to return the castaways to their homeland.

But when the *Manhattan* made her landfall, and Captain Cooper sent messengers ashore requesting permission to anchor in Uraga Bay, off the capital city of Edo (Tokyo), the suspicious Japanese initially wanted to drive the "big junk" away without even landing the men it had saved. A few officials, however, were moved by the evident hu-

manity of the *Manhattan*'s mission. And after three weeks of discussion, the *Manhattan* was at last allowed to drop anchor off Edo. There, to the surprise of the whalemen, the Japanese immediately surrounded the whaler with 500 guard boats.

Shortly, a delegation of silk-robed dignitaries arrived, accompanied by an interpreter and artists with fine-tipped brushes, inkstones and ample supplies of rice paper. It soon became evident that the interpreter did not understand English, but he delivered one message very explicitly in sign language—drawing a naked sword across his throat to indicate the penalty for leaving the ship.

That established, the Japanese politely asked about the ship's needs for provisions, and arranged the details of their nationals' repatriation. When the business was finished, the Japanese dignitaries enthusiastically consented to a tour of the ship.

As the artists sketched furiously, recording every detail, Captain Cooper explained the use of the sextant, and played a tune on the organ in his cabin. However, what intrigued the Japanese

most, according to an observer, were the eight crew members "with faces and bodies as dark as charcoal"—the Japanese apparently had never seen black men before.

After its visit the delegation offered gifts of lacquer bowls and local delicacies such as octopus, as well as more standard provisions. As a special token of gratitude and esteem, the Emperor sent Cooper his imperial autograph—which looked to one of the Americans "as if a half-grown chicken had stepped into muddy water and then walked two or three times deliberately over a sheet of coarse paper."

Amid much bowing and ceremonial leave-taking, the rescued Japanese finally departed the *Manhattan* on April 20. Captain Cooper reported that a number of them actually wept on bidding goodbye to their saviors. The whaler weighed anchor the next morning, and as she did so, the entire armada of 500 guard ships formed into ranks and, attaching themselves to her bow, towed her out to sea. It was, remarked Captain Cooper, with typical New England restraint, a farewell gesture "approaching the marvelous."

life with an explosion of energy. He smashed one of the *Crieff*'s boats and, still dragging the *Yankee*'s boat, charged at one from the *Dudley*. Missing on the first try, he spun and seized the craft in his jaws, shaking it apart and actually gulping down two of its occupants. The men in the *Yankee*'s boat cut their line and, with the other crews who were still afloat, pulled for the *Crieff*, which had finally arrived on the calamitous scene. As they clambered up the ship's side, Mocha Dick turned away to ram the now-deserted hulk of the lumber schooner. But their ordeal was not over. Suddenly the whale breached from beneath the *Crieff*, grazing her bow and taking away the bowsprit and jib boom. Finally the whale swam away, turning to watch placidly as the crippled *Crieff* retreated.

There are several versions of how Mocha Dick died. The most widely accepted is that he was killed by a Swedish whaler that brought the Leviathan to bay in 1859; he was blind in one eye and evidently too tired to struggle as the lance punctured his lungs. Mocha Dick carried 19 irons in his scarred hide—mementos of more than 100 battles in which at least 30 men had been killed and scores of whaleboats destroyed. It was inevitable that his story would pass into literature, and that this old rogue would become the model for Moby Dick, central figure of one of the greatest sea stories—and morality tales—ever written.

By every account, Mocha Dick was endowed not only with prodigious strength but with a malign intelligence as well: he feigned death; he stood vigil after a fellow whale was killed; he even, as one observer put it, seemed "particularly careful as to the portion of his body which he exposed to the approach of the harpooner." In these combined qualities of power and cunning, Dick became the supreme representation of traits that whalemen, respectful of their formidable foe, attributed even to lesser members of the species.

There could be no doubt whatever about the brute force a whale could exert—or its stamina. In 1817 a right whale pulled six of the *Royal Bounty*'s boats and 1,600 fathoms of line across the ocean for more than 24 hours; when the men succeeded in passing the line to the ship, the whale dragged the *Royal Bounty* along at two knots for one and a half hours before finally tiring and giving up.

At about the same time, the *Winston*'s captain, Edmund Gardner, became one of the few men ever to survive the crushing power of a sperm whale's jaws. While stalking a whale off the west coast of South America, Gardner went forward in his whaleboat to assist the harpooner. He planted a harpoon and was reaching for a second one when the whale struck. Later Gardner recalled only the flashing of huge teeth as they closed on him. His account continued:

"When I came to my senses after being stunned, I called one of the boat's company to cut off the line and take me to the ship. I was bleeding copiously when taken on board. My shoes were quite full of blood. When on board, found one tooth had entered my head breaking in my skull, another had pierced my hand, another had entered the upper part of my right arm, the fourth had entered my right shoulder, from the shoulder to the elbow of the right arm was badly fractured. My shoulder was broken down an inch or more (where it now is), my jaw and five teeth were

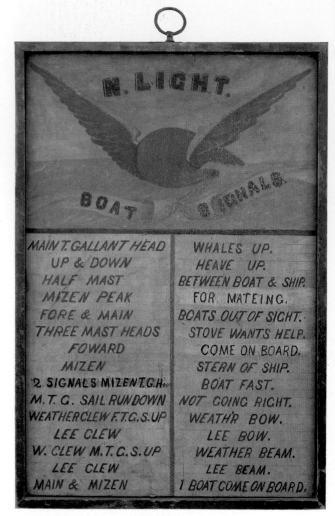

A decorative plaque displays the signals the whaleship Northern Light used to pass information to her whaleboats. The first nine signals and the last were made by flying flags at the mastheads. "Heave up" meant to hold the boats where they were; "for mateing" meant to join another whaleboat in pursuit of a whale. The other five signals were made by clewing—i.e., partially shortening—sails.

broken, tongue cut through, my left hand was pierced with a tooth and much broken and very painful." Captain Gardner was put ashore at Paita, Peru, to recover from his wounds. He recuperated enough to go whaling on many more voyages and live to the ripe age of 90.

As for the intelligence of the sperm whale, Owen Chase of Nantucket was convinced to the end of his days that the most famous whaling disaster in history—the wreck of the *Essex*—had been the result of a sperm whale's "decided, calculated mischief." Chase was in a good position to render this judgment: he was first mate of the *Essex* during its final, fatal affray.

On November 20, 1820, the ship's boats were chasing whales in an open stretch of the Pacific 1,200 miles northeast of the Marquesas Islands. Chase drove a harpoon into one whale, but a swipe of its flukes stove in his boat. Cutting the line and stuffing shirts in the hole, the men rowed back to the ship. There Chase was supervising repairs on the boat when he looked up and, to his horror, saw an enormous sperm whale—Chase guessed its length at 85 feet—heading hell-bent for the *Essex*: "He came down on us with full speed and struck the ship with his head, just forward of the fore-chains. The ship brought up as suddenly and violently as if she had struck a rock and trembled for a few seconds like a leaf. The whale passed under the ship, grazing her keel as he went along and came up to leeward, apparently stunned with the violence of the blow."

The *Essex* almost immediately began settling in the water, but the whale was by no means done with the stricken ship, fetching her another terrible blow in the bow. To Owen Chase, the attacks could have come only from "premeditated violence." He even attributed to the whale a knowledge of elementary physics: the blows, he wrote, "were calculated to do us the most injury. By being made ahead, they thereby combined the speed of the two objects for the shock. To effect this impact the exact maneuvers which he made were necessary."

Within 10 minutes the decks of the *Essex* were awash. And when Captain George Pollard and Second Mate Matthew Joy, who had all the while been out in their boats unsuccessfully trying to kill whales, finally rowed back toward the *Essex*, they were astonished to see her masts roll over into the water. Pollard's boat was the first to reach Chase, who had managed to launch a boat and was a few ship lengths away from the sinking vessel.

"My God, Mr. Chase," Pollard asked, "what is the matter?"

"We have been stove by a whale."

Buoyed by air trapped in her hull, the *Essex* stayed afloat long enough for the men to supply the whaleboats with meager provisions. Then the three boats, carrying a total of 20 men, set out across the Pacific for the South American coast.

Day after day the pitiless Pacific sun beat down on the wretched castaways. They suffered from salt-water boils. Their lips cracked and bled. Starvation cramps wrenched their bowels. On the 51st day, in Chase's boat, one man died. Two days later a squall provided precious fresh water—but separated the bobbing boats for the rest of the journey.

Another man in Chase's boat died and, on the 81st day, a third. Only three days' supply of food remained, and Chase himself made the sug-

gestion that would haunt the survivors for the rest of their lives. "We separated his limbs from his body," Chase wrote, "and cut all the flesh from the bones, after which we opened up the body, took out the heart, then closed it again—sewing it up as decently as we could—and committed it to the sea."

While their shipmate's carcass fed the sharks, his heart and some of his flesh kept the rest of the crew alive—until, after drifting 4,500 miles across the Pacific for three months, the whaleboat and its scarecrow crew were sighted and rescued by the brig Indian of London.

In Captain Pollard's boat, starvation drove the men to an even more excruciating decision: they drew lots to select a victim to be killed in order that the others might live.

The lot fell to young Owen Coffin. He was Captain Pollard's nephew, and Pollard offered to take his place, but the lad would not have it. The victim selected his own place of execution. As Pollard later described the scene, "The poor emaciated boy hesitated a moment or two; then, quietly laying his head upon the gunwale of the boat, he said, 'I like it as

A wounded and vengeful sperm whale flips a whaleboat skyward with its great flukes after staving in the hull of the ill-fated whaler Essex, in this 19th Century re-creation of the tragedy that occurred in the Pacific on November 20, 1820. One survivor, First Mate Owen Chase, recalled how the maddened bull gave the whaler "such an appalling and tremendous jar, as nearly threw us on our faces."

well as any other.' He was soon dispatched and nothing of him left. But I can tell no more—my head is on fire at the recollection."

By February 23, 1821, after 96 days on the water, Pollard's whaleboat was in sight of Santa Maria Island, off the coast of Chile, when the survivors were rescued by the Nantucket whaler *Dauphin.* The third boat was never seen again. Of the 20 crewmen of the *Essex*, only eight survived. Five of them went right back to whaling—and all five, including Owen Chase, later became captains.

As a young man serving in the fo'c's'le of the Fairhaven whaler *Acushnet*, Herman Melville read Owen Chase's published account of the *Essex* disaster; inevitably, Melville also heard countless tales about Mocha Dick. His great American novel, *Moby-Dick, or the Whale*, which combined and embellished the stories of the *Essex* and Mocha Dick, was published in November 1851. That month, the author heard accounts of another ship-killing whale that made him wonder "if my evil art has raised this monster."

The victim was none other than the *Ann Alexander*, whose captain, John DeBlois, had on this same voyage already suffered considerable indignity because of a "knowing" whale, as he termed it in his letter to his wife. The *Ann Alexander* had not been designed as a whaler. She was built in South Dartmouth as a merchantman and was launched in 1805 to the invocation of a Quaker lady: "I christen thee *Ann Alexander*, little ship, and may thee ever sail swiftly and safely to thy desired ports." After the Pacific grounds opened and profits from whaling rose, the *Ann Alexander* was converted and enlarged to provide more cargo space. She had been whaling for almost 30 years and had captured hundreds of whales before she met her doom in the Pacific.

As was later related in a newspaper account, the ship was on the Offshore Grounds when, late one afternoon, lookouts sighted a whale. Two boats were lowered but returned to the *Ann Alexander* when darkness made further searching impossible. DeBlois accurately calculated the course and speed of the whale and pursued it throughout the night, coming up on the quarry the following morning. Boats were put down, and as the mate's craft got close to the whale, DeBlois sang out, "This is a noble fellow. Don't galley him." Despite the warning to the men, the whale sounded, and DeBlois figured "from the way he turned his flukes that he'd go fully three miles under water." The boats hurried to meet him and, when the whale surfaced between the boats, the mate struck him with a harpoon.

The wounded whale rushed the mate's boat and missed, then spotted the captain's boat and, DeBlois was quoted as saying, "started thoroughly in a rage for me. He reared up so perpendicularly that he lost his headway. Baffled, he settled down in the sea." Astoundingly, the whale had shot so far out of the water that it had lost its balance and flopped over on its back. And now, somehow "our boat was grounded on his belly, no very gentle resting place for a frail boat! Thus for a moment we lay aground on the mighty carcass. Then the monster straightened out and shot ahead, leaving us afloat."

Next, the whale attacked the mate's boat, DeBlois told the reporter, "and in an instant it was crushed like so much paper between his mighty jaws. The men were thrown hither and thither and, climbing on the broken boat, were again dashed from it. Two men were thrown fully 20 feet in the air by a vicious lunge of the whale."

The second mate had remained with the *Ann Alexander*. But now he lowered a boat to give assistance. As DeBlois and the second mate fished their comrades from the seas, they discovered that the line was still attached to the whale, and made it fast to DeBlois' craft. The other boat, being lighter and faster than the captain's boat, set off in pursuit of the whale in order to dart a second harpoon. But when the whale attacked and destroyed that craft, DeBlois cut the line, picked up the swimming men and headed for the ship "without delay."

Back aboard the *Ann Alexander*, the ever-persevering captain decided to give chase in the ship. Lance in hand, he was standing on the bow when the whale "struck the ship with a dull thud which knocked me off the bow clean on the deck." To DeBlois' relief, the *Ann Alexander* still seemed sound—and he was still determined to get the whale. But when he attempted to lower a boat, the men, who had clearly seen enough of this whale, resolutely refused to move. Even as they were arguing, the whale returned to the attack. "I caught a glimpse of a shadow," said DeBlois, "when the whale again struck the ship a terrible blow which shook her from stem to stern. The destroying monster had hurled himself against the bow four feet from the keel. I at once attempted to go down into the forecastle, but heard the water rushing in at a rate that I knew it was hopeless."

Abandoning the broken *Ann Alexander*, captain and crew put out in their boats; after nearly two weeks they were picked up by the whaler *Nantucket*. And about five months later the *Rebecca Simms* of New Bedford killed an old, weary whale that failed to put up a fight. Fast in his carcass was an iron from the *Ann Alexander*—and splinters from ship's timbers were deeply embedded in his head.

To the Yankee whalemen the Atlantic was "this side of land" and the Pacific was "the other side"—and when they first went there it was a world apart from anything they had known. The islands, of course, were endlessly beguiling. In the early 1800s the girls from the Sandwich Islands swam naked to the ships and climbed aboard to trade themselves for trinkets. They were not prostitutes for the simple reason that they were breaking no moral code they knew. One whaleman recorded the purchase of pigs and chickens and added that "other creatures of a more alluring and captivating kind presented themselves in a row."

The Pacific islanders had seen scarcely any white men since Cook's 18th Century explorations, and they were as fascinated by the pale-skinned and heavily clothed visitors as the latter were by their tattooed, nearly naked hosts. These early contacts were invariably cordial and often were accompanied by ceremonial feasts that would have scandalized the strait-laced New Englanders back home.

Captain Richard Macy of the *Maro* was greeted in the 1820s by a Fijian chief and his retinue, "in a state of nudity, with the exception of a little grass." Macy went ashore with "the king," as he called the chief. "The

The odyssey of an American Homer

A PORTRAIT OF HERMAN MELVILLE, PAINTED ABOUT 1870

The first whales Herman Melville ever saw, as a 19-year-old seaman on board a trans-Atlantic packet in 1839, were a "bitter disappointment." They were so small, he wrote, that "I lost all respect for whales." The cetaceans in question were probably pilot whales, a species that rarely grows longer than 20 feet. Once Melville had seen other examples he quickly regained his regard for the order; within 18 months he had signed on the New Bedford whaler *Acushnet*, lured to sea by what he called "the overwhelming idea of the great whale himself."

The son of a once-wealthy New York family impoverished by the depression of 1837, Melville was about to experience an odyssey in the Pacific as dramatic as anything that later came from his pen. After 18 months of grueling work and greasy luck, the *Acushnet* anchored in the harbor of Nuku Hiva in the Marquesas Islands. The young whaleman watched the surf curling on the white beach and the naked island girls swimming out to the ship—and immediately deserted. The islanders welcomed Melville as a curiosity. Servants fed him exotic foods, and island girls swam with him in the lagoon. But as a pampered pet he was kept under guard, and only with difficulty did he escape a month and a half later to sign aboard a whaleship from Australia, the *Lucy Ann.*

The *Lucy Ann* made the *Acushnet* seem like a pleasure yacht. Her forecastle swarmed with rats and her ailing captain left the crew to the mercy of a bullying, drunken mate. As might be expected, Melville and some of his shipmates refused to work, stopping just short of mutiny. When the ship reached Tahiti, they found themselves in the wooden stocks of the local jail, the Calabooza Beretanee. But the easygoing Tahitians released Melville and his shipmates as soon as the *Lucy Ann* sailed.

For almost a month Melville roamed the beaches of Tahiti and nearby Moorea. As on Nuku Hiva, he got along famously with the islanders. Nevertheless, by now Melville was beginning to yearn for civilization, and on November 7, 1842, he took passage on the Nantucket whaler *Charles and Henry* for Honolulu. From there, after a spell, he headed home: the general muster roll of the frigate *United States* for August 17, 1843, included Ordinary Seaman Herman Melville.

The *United States* cruised slowly through the Pacific and around the Horn, and it was not until October 3, 1844, that Melville stepped onto a Boston pier, after nearly four years of heady adventure. He shortly put his experience to work, writing five successful novels about seafaring and life in the South Seas. Then, at last, he was ready for his masterpiece.

"On the hither side of Pittsfield," wrote Melville's neighbor Nathaniel Hawthorne in 1851, "sits Herman Melville, shaping out the gigantic conception of his white whale." *Moby Dick*—Melville's almost Shakespearean tragedy of the quarter-deck King Lear, Captain Ahab, and his thundering adversary, the rogue whale Moby Dick—eventually was to win acclaim as the most moving and realistic account of Yankee whaling ever produced. But it received no such recognition in the author's lifetime. One critic called it "the worst school of Bedlam literature."

Melville, 32 at the time *Moby Dick* was published, lived 40 more years in increasing obscurity. When his death was briefly reported in *The New York Daily Tribune* on September 29, 1891, the paper identified him as an author. Reading the obituary, the editors of *The Critic,* the leading U.S. literary magazine, were at a loss. They had never heard of Herman Melville, or of *Moby Dick.*

king introduced me to the queen, who was apparently much pleased to see me," wrote Macy. "I was seated on a clean mat and fanned by a woman on each side of me. The queen spread a table, which was a large wooden tray spread with leaves; and the meal consisted of yams, breadfruit, tarrow, fish, cocoanuts and other dishes, which were prepared under the immediate inspection of the queen. She handed me each dish separately in a leaf, taking care not to touch her fingers to either."

Some sailors were less fortunate in their feasts. The New Caledonians enjoyed bats, tree grubs and pigeon entrails in rice. That intrepid harpooneer, Nelson Haley, described a dinner on Strong Island, in the Carolines, during which he watched the islanders chew kava root, spit it into a calabash shell and pass it around for the guests to drink. Haley happened to glance at a shipmate: "just as he removed the shell that contained the dose he had been drinking away from his mouth, I saw strings of the vile decoction showing in the bright light, running down his moustache. I left the table pretty suddenly."

Inevitably the men from around the Horn altered the islanders' way of life. In this they were abetted by adventurers and entrepreneurs from Europe as well as America, chiefly British jailbirds who had escaped from the penal colonies in Australia. Within a few years of the whalemen's arrival in the Sandwich Islands in 1819, the swimming wahines were augmented by such establishments as Murphy's grog shop, where girls danced nude, and houses of ill fame. Then came the missionaries.

Some whaling captains welcomed these pious, teetotaling newcomers. A number of masters kept what were called "temperance ships," banning all liquor to the crew—though of course they carried a supply for trading with the islanders. But the missionaries carried matters a large step further with actions that puzzled the islanders—and for the most part outraged the whaleship hands. At first it amused the whalemen to see the effects of the missionaries' attempts to clothe the islanders. Derby-hatted chiefs were a frequent source of merriment, and one whaleman enjoyed recalling the spectacle of a dignified islander in a beaver hat, a shirt and nothing else. Another Yankee was impressed by a Tahitian wearing a Western-style coat and pantaloons, the six-inch gap at his midriff decorated by a silk handkerchief tied in a bow.

But nothing could compensate the whalemen for the changes wrought by the missionaries in the islanders' moral behavior. Missionaries banned the famous hula and even the wearing of flowers after they discovered that the way a girl wore a flower in her hair could constitute an invitation. Whalemen were especially incensed to find that local officials in Honolulu, acting at the behest of the missionaries, were levying fines on the sailors for what shortly before had been acceptable behavior. Some of the penalties: noisemaking, one to five dollars; speeding on horseback, five dollars; desecrating the Sabbath, six dollars; drunkenness, six dollars; fornication, five dollars; rape, six dollars; adultery (or, because the Hawaiian language had no word for it, "mischievous sleeping"), thirty dollars.

Many islanders adapted to the new moral code in their own ways. A Tahitian girl earnestly assured Melville that she was "mickonaree" in head and heart but not elsewhere. However, the churchmen were persistent and generally prevailed. The Reverend Henry T. Cheever, censuring the thriving, wide-open port of Lahaina as "one of the breathing holes of hell," persuaded the town fathers to deny the whalemen their daughters. When the Lahaina bawdyhouses were shuttered, whalemen rioted. In Honolulu, reacting to alleged police brutality in enforcing the new codes, whalemen drank and looted their way through the town, eventually setting fire to a police station and nearly burning Honolulu's waterfront. A fortunate shift of wind quelled the conflagration, and a citizens' guard of some 500 armed men, mustered by residents and shipmasters, broke up the rioters and restored order. In the Friendly Islands, Nelson Haley reported the consternation of the *Morgan*'s fourth mate when an island girl attempted with suggestive sign language to sell him some bananas. The mate slipped an arm around her waist—only to have her run off and admonish him in English, "Devil! Devil! Too much Devil!"

The fact is that, however the islanders dressed or behaved, the whalemen's influence on them was hardly called beneficial. To begin with, the sailors spread influenza, tuberculosis, cholera and other diseases among the islanders. The impact could be catastrophic. The islanders had no hereditary resistance to such mild white man's maladies as measles and chicken pox, which devastated whole settlements. In the 1770s Tahiti's population was about 40,000; by the 1830s diseases had reduced it to 9,000. Nearly 30,000 Fijians died in one measles epidemic in 1875.

The innocent islanders were often ill-treated by the visitors. Some of the less principled whaling captains "paid with the foretopsail," agreeing to buy native produce and then sailing away as soon as their decks were full. Others even shot their benefactors. Such an incident need occur only once for an island's inhabitants to start killing all white visitors, even those "with salt water in their eyes," i.e., deserters and shipwrecked sailors swimming ashore.

Eventually some of the islanders themselves became practiced in deceit, pretending to be friendly until they could overpower and massacre the whalemen. When the *Awashonks* of Falmouth touched at Namarik Island in the Marshalls in 1835, a group of islanders came aboard to trade, showing no signs of animosity until a moment when most of the crew was preoccupied. Then at a prearranged signal, the visitors made a rush for the ship's cutting spades, with which they beheaded Captain Prince Coffin and killed the helmsman and both mates.

The rest of the crew barricaded themselves belowdecks, while the islanders rampaged around overhead. At last the third mate had an idea. He lodged a keg of gunpowder at the top of the companionway, with a trail of powder on the steps. Touching off the powder trail, the men below waited until it sizzled up the companionway to the keg. Immediately after the explosion the sailors burst through the smoke and drove off the terrified islanders. The third mate thereupon took the *Awashonks* away from Namarik Island, as far—and as fast—as he could.

Hostile inhabitants of the Palau Islands were thwarted by another ingenious tactic. Swarming aboard the whaleship *Syren*, they forced the crew to flee into the rigging. Perched on high, one of the men found a box of tacks that someone had left in the maintop. He scattered them on the deck, with an instantaneous effect on the islanders' bare feet. Howling

The "Lagoda": stout, sturdy and capacious

One of the most successful whalers ever to ply the oceans was not even built for her profession. When the *Lagoda* was sold to a New Bedford whaling concern in 1841, she already had 15 years' service as a Boston merchantman behind her. But she had all the attributes demanded of a good whaler.

She was framed and planked throughout with durable oak. Her bulky, tub-shaped hull was right for whaling, in which speed meant little and the best ship was the one that could plod along month after month accumulating oil in her capacious hold. The 371-ton *Lagoda* made little more than five knots.

At New Bedford she was fitted out for her new role with five long, slim whaleboats hung from davits off her rails, two on the starboard side and three on the port. Her rigging was changed from that of a full-rigged ship to that of a bark, with fore-and-aft sails on her mizzenmast. This rig was preferred on whalers; because no one was needed aloft to handle the mizzen, more hands were free to man the boats. A white horizontal stripe broken by black squares was painted on her side. The squares looked like gunports at a distance and were intended to deter pirates or hostile savages.

Her first voyage yielded a fabulous 2,700 barrels of oil and 17,000 pounds of baleen. And under a succession of captains, she served for almost 50 years, landing an unrivaled 31,409 barrels of oil and 267,058 pounds of baleen. At last in 1890, a North Pacific gale damaged her so badly that she barely reached Yokohama, where she ended her days as a coal hulk fueling steamers in Japanese waters.

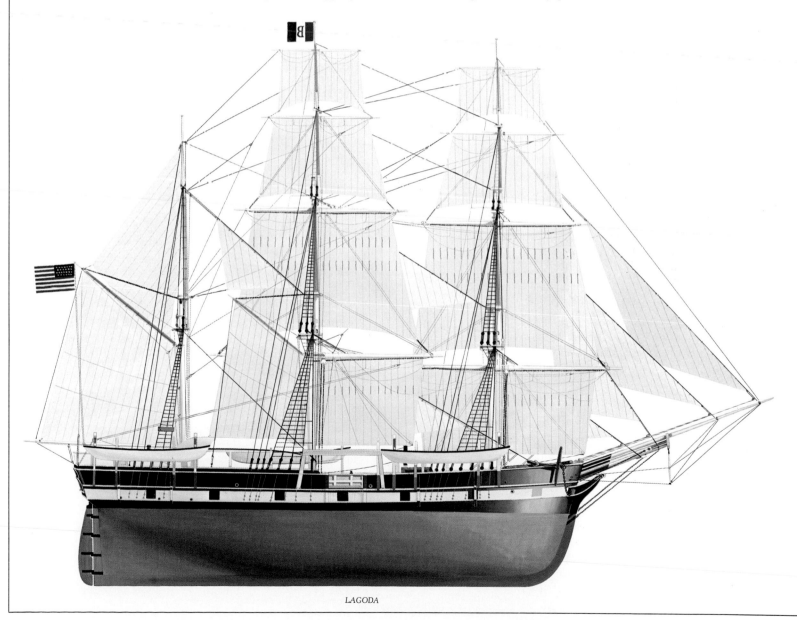

LAGODA

At 108 feet, with a beam one-quarter her length and a depth of 18 feet, the *Lagoda* was of average size for a whaler. In this space she managed to accommodate a crew of 30 officers and men and the provisions and equipment necessary for her complicated business. But it was a tight squeeze.

At her stern the afterhouse covered the steering mechanism, which whalemen aptly called the shin breaker. The wheel was mounted on the end of a heavy tiller that moved in an arc a few inches above the deck. As the wheel was turned, it pulled the tiller to a new position by a system of ropes and pulleys. It took an agile helmsman to avoid a knock or two in heavy weather as the assembly of tiller and wheel swung to and fro across the deck. In a skylight just forward of the wheel was the ship's compass, a two-sided device that could be read by both the helmsman on deck and the captain at his desk below.

The captain's quarters boasted the only suggestion of elegance on board. The bed swung on gimbals to counteract the roll of the ship, and there was a desk, chair and sofa. The mates slept in tiny staterooms off the main cabin, and the ordinary whalemen bunked in the fo'c's'le. The men entered through a scuttle from the main deck just aft of a windlass used for hoisting blubber cut from the whale.

Between the fo'c's'le and the steerage was the blubber room, where the blubber was chopped up for the tryworks, a heavy brick and cast-iron stove anchored to the main deck aft of the foremast. A few casks of oil were stored in the blubber room, but most were kept below in the hold.

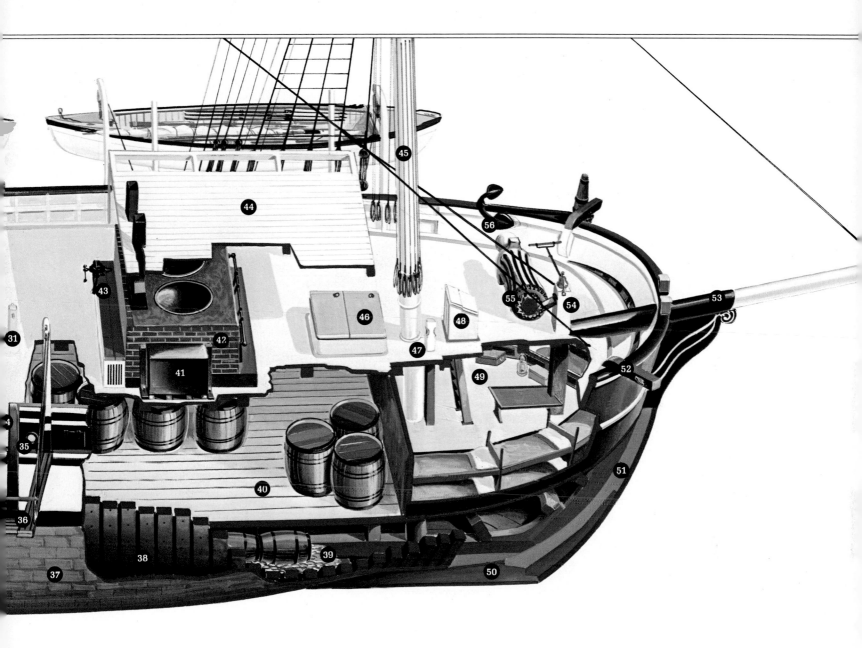

1. WHALEBOAT	15. RUNNING LIGHT	29. FIFE RAIL	43. WORKBENCH
2. AFTERHOUSE	16. SPARE WHALEBOAT	30. MAIN HATCH	44. TRYWORKS SHELTER
3. PAINT LOCKER	17. MIDSHIP SHELTER	31. DECK BITT	45. FOREMAST
4. COMPANIONWAY	18. SECOND MATE'S CABIN	32. DECK BEAM	46. FOREHATCH
5. SKYLIGHT	19. PANTRY	33. HATCH TO HOLD	47. FLUKE BITT
6. COMPASS	20. FIRST MATE'S CABIN	34. GANGWAY	48. FO'C'S'LE COMPANIONWAY
7. WHEEL	21. HOLD	35. SCUPPER	49. FO'C'S'LE
8. TILLER	22. FALSE GUNPORT	36. CUTTING-IN STAGE	50. KEEL
9. CAPTAIN'S CABIN	23. STEERAGE	37. COPPER SHEATHING	51. STEM
10. CAPTAIN'S HEAD	24. STEERAGE COMPANIONWAY	38. FRAMES	52. CATHEAD
11. RUDDER	25. PINRAIL	39. BALLAST	53. BOWSPRIT
12. CAPTAIN'S STATEROOM	26. MAINMAST	40. BLUBBER ROOM	54. SHIP'S BELL
13. MAIN CABIN	27. DAVIT	41. COOLING TANK	55. WINDLASS
14. MIZZENMAST	28. GRINDSTONE	42. TRYWORKS	56. ANCHOR

The Yankee whaleboat was a light, swift, superbly nimble craft typically about 28 feet long and six feet wide. It was double-ended so it could reverse directions easily, and was planked with rot-resistant cedar only half an inch thick.

Hanging on davits, the boat weighed about 1,000 pounds, but in the water it could hold another 1,000 pounds of gear plus the combined weight of six whalemen. Pulling on oars 14 to 18 feet long, five men could propel the craft at a speed of five knots, while the mate at the stern could turn her in a wink with a twist of the 20-foot steering oar.

The whale line, coiled in tubs near the centerboard, was passed around a loggerhead at the stern and then strung forward through a chock at the bow, where it could be deftly paid out in the wake of a rampaging whale.

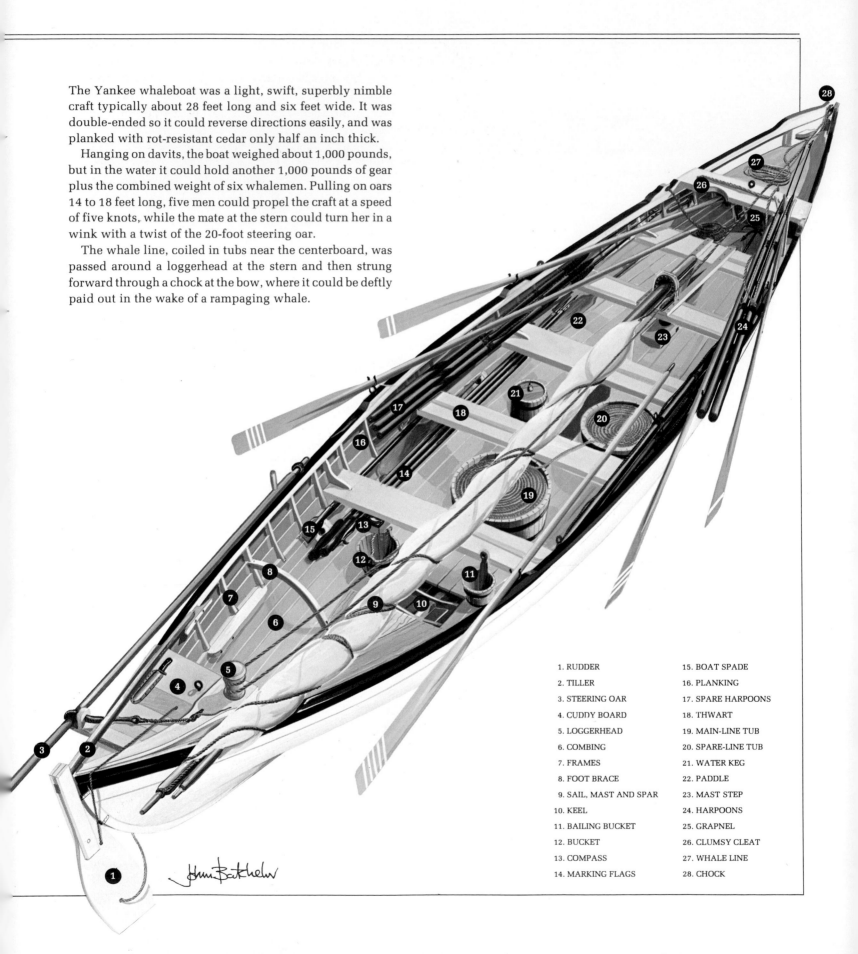

1. RUDDER	15. BOAT SPADE
2. TILLER	16. PLANKING
3. STEERING OAR	17. SPARE HARPOONS
4. CUDDY BOARD	18. THWART
5. LOGGERHEAD	19. MAIN-LINE TUB
6. COMBING	20. SPARE-LINE TUB
7. FRAMES	21. WATER KEG
8. FOOT BRACE	22. PADDLE
9. SAIL, MAST AND SPAR	23. MAST STEP
10. KEEL	24. HARPOONS
11. BAILING BUCKET	25. GRAPNEL
12. BUCKET	26. CLUMSY CLEAT
13. COMPASS	27. WHALE LINE
14. MARKING FLAGS	28. CHOCK

with pain, they all jumped overboard, leaving the *Syren* to sail away.

In dealing with the islanders, the whalemen never knew quite what to expect, and after a time were constantly on their guard. Some of the most convincing testimony to this general rule was offered by Nantucket whaleman William Cary, who spent more than two years of involuntary residence in the Fiji Islands.

Cary was a foremasthand aboard the whaler *Oeno* when she was wrecked on a reef off Turtle Island, one of the Fijis, in 1825. Captain Samuel Riddell and his crew of 20 men had no choice but to seek shelter with the Turtle Islanders, and Riddell at first ingratiated himself with his hosts. But presently another less friendly tribe arrived in war canoes, and the atmosphere became menacing. Finally, when an *Oeno* crew member rapped the knuckles of an islander trying to eat from the whaleman's dish, William Cary sensed that the tension had reached a bursting point, and he did not wait for the explosion.

Crossing to the other side of the island, Cary found a cave where he hid for the night. Next morning he crept out and went around the tip of the island, keeping to the beach. At the landing where the *Oeno*'s men had come ashore he found no one. "I searched around," he later wrote, "until I found a place which evidently had been dug over. I scooped away a few inches of sand with my hand and came to the face of a man." It was soon obvious to Cary that every one of the *Oeno*'s crew had been murdered.

Cary returned to his cave, but after three days hunger persuaded him to give up. When a native hunting party neared his hiding place, he crawled out again. "They immediately started for me," Cary wrote, "one armed with a boat hatchet, the other with a knife. I sat down in the path with my back towards them, expecting to have the hatchet driven into my head, and not wishing to see the blow. They walked up until within a few feet of me, then stopped and looked at me a moment before they spoke. It seemed an hour."

With the party, as it turned out, was an old man—a tribal chief—whom Cary had previously befriended. The patriarch introduced Cary to the visiting tribesmen as his adopted son. His foster father's generous act no doubt saved Cary from torture and death. The blessing was mixed: Cary was still alive—but only as a sort of pampered prisoner.

During the many months that followed, William Cary wore nothing save an island skirt. He learned to speak the islanders' language and to drink the sickening, potent kava. He went diving for *béche-de-mer*, the sea slug that white traders bought to sell in China, where it was a prized delicacy. When one of the chiefs was killed by a usurping brother, Cary attended the elaborate funeral ceremony, at which the chief's four widows were strangled so that they might accompany him to the hereafter. Next day Cary watched the mourning ritual, in which the chief's subjects, men and women, showed their grief by shaving their heads and amputating the little fingers of their children.

The Fiji tribes were forever fighting with one another, so Cary's knowledge of firearms made him popular with the chiefs and was his best protection. He repaired their rusty muskets and he even went along on several war parties. On one of those excursions Cary related that the 4,000-man force took and burned seven villages in one day. The warriors

cleared the area of enemies and then, Cary wrote, "we marched down to our canoes, taking with us five or six female prisoners and some of the dead bodies for a cannibal feast after we got home."

For Cary, the route to freedom opened when a long-haired visitor from another island, white but leathered by the sun and dressed in an islander's skirt, walked up to Cary, called him by name and asked if he remembered David Whippey. The puzzled castaway replied, "Yes, I formerly knew him. He was a townsman of mine and an old playmate."

"Well," said the newcomer, "I am that David Whippey."

William Cary's former Nantucket playmate had shipped out as a whaleman aboard the *Francis* in 1818, some nine years before. But upon deciding that he did not like whaling, Whippey had deserted. He and the ruler of Ambow Island, one Kakombau, took a liking to each other, and Kakombau established the American as lord of nearby Ovalau Island. Whippey now used his influence with the tribesmen to win Cary's release. As for himself, he remained behind on his tropical island and was eventually appointed U.S. vice consul in the Fijis. He also took several wives and established a populous Whippey dynasty in the islands.

The South Pacific was filled with men who, like David Whippey, had fled the dangers—and the tedium—of the whaleships. One U.S. consul in the Pacific estimated that 3,000 to 4,000 Americans deserted their whaleships every year. For many, of course, the allures to be found on the palm-fringed beaches were too great to resist. But others were simply seeking refuge from service under harsh masters.

Fashioned from the feathers of rare iiwi birds, this splendid cape was presented to Captain Valentine Starbuck of the whaler L'Aigle in 1824 by the secretary to Hawaii's King Kamehameha II.

Whalemen were sometimes flogged, though less often than navy men. Whaleship punishment also included bread-and-water confinement in the run—a dark, airless compartment belowdecks. In the equatorial heat along the line, this could be worse than flogging. And one poor wretch had to spend every waking hour for a month endlessly scouring the try-pot with ashes. Such treatment not only drove men to jump ship, it could push them to the edge of mutiny—and sometimes beyond.

Captain Thomas Worth was unquestionably a hard man. He had assumed command of the *Globe*, the vessel in which George Washington Gardner had discovered the Pacific's Offshore Grounds, when Captain Gardner accepted command of a larger ship and recommended Worth for the position. The new master had served as first mate on the *Globe*'s previous voyage and had earned a reputation as a stern taskmaster.

In 1824, while on a cruise to the Pacific, Worth was the victim of the bloodiest mutiny in Yankee whaling history. The trouble started when the vessel was provisioning in Honolulu and six of her crew deserted. They were replaced by the best men Worth could find—derelicts from the waterfront who quickly joined an existing cabal of malcontents led by one Samuel Comstock, a harpooneer with a record for troublemaking. The vessel had been at sea for 28 days when one of the new hands, a seaman named Joseph Thomas, became so insubordinate that Captain Worth had him triced to the rigging for 15 lashes. As was customary, the crew was required to watch. The spectacle so angered a number of men that they readily joined Comstock when he proposed mutiny.

In one gory night Comstock and his followers murdered the captain and all three officers, Comstock himself dispatching the sleeping Worth by decapitating him with an ax. Tossing the officers' bodies overboard (one of them mortally wounded but still alive), Comstock took the *Globe* to the Mulgrave Islands, where he planned to set up a fiefdom enforced by the ship's armaments.

Perhaps inevitably, the mutineers soon fell to murderous brawling. Comstock was shot to death and half a dozen survivors managed to slip the *Globe*'s cables and escape to sea, arriving in Valparaiso four months later. They were questioned by the U.S. consul and sent home to Nantucket; because of various extenuating circumstances, only one man ever went to trial, and he was acquitted of mutiny.

The mutineers left on the island so mistreated their native hosts that all but two were slaughtered in a retributive massacre. These two marooned survivors, teenagers Cyrus Hussey of Nantucket and William Lay of Saybrook, Connecticut, had been forced by threats to join the mutiny. They now owed their lives to the fact that they had been friendly toward some of the Mulgrave Islanders, who saved them from the massacre. They were adopted by the tribe and lived for nearly two years in the Mulgraves, until a U.S. Navy force picked them up.

Sometimes a renegade sailor, because of his firearms, would assume command of a tribe and lead it in vicious attacks on visiting whaleships. When the whaler *Triton* made a landfall off Sydenham Island in the Kingsmill Group in 1848, tribesmen led by a Portuguese deserter named Manuel lured the captain ashore and imprisoned him. At dawn the following day they went out to the ship and attacked the crew.

Honolulu's Sailors' Home and Bethel flies a welcoming pennant for the thousands of Yankee whalemen who stopped off at the tropical Sandwich Islands in the 1850s and 1860s. "Could I have forgotten the circumstances of my visit," remembered a visiting sailor, "I should have fancied myself in New England."

The "Oscar" case: mutiny or murder?

An ax-waving mutineer is slain on board the Oscar, after which his body is ferried ashore (right) in this composite of the 1845 affair.

"You ought to have taken the boat and gone after them, and run a lance through them!" bellowed Captain Isaac Ludlow at his first mate as they stood peering at the splashing forms in the distance. It was not whales the captain of the *Oscar* yearned to lance. His rage, that August day in 1845, was directed at three of his own seamen who had defied orders by swimming ashore for unauthorized liberty.

Thus began one of the most bitterly debated incidents of conflict at sea in the annals of American whaling. Before the events had run their course, they threatened to topple that most sacred tenet of maritime law—the absolute authority of the captain.

The *Oscar*, 291 days out from Sag Harbor, New York, was provisioning at Ilha Grande, Brazil, when the three sailors paddled ashore without leave. The men stayed only long enough to get roaring drunk, and then swam back to the ship, where they were sent below while the captain pondered a proper punishment.

Grumbling and cursing, the besotted sailors armed themselves with knives, axes and cudgels. Joining them was the cook, who believed himself aggrieved on another score. The captain met them with a rifle. He warned that he would shoot the first man who set foot on the quarterdeck. When a seaman lunged forward, the captain shot him dead.

The surviving seamen were immediately placed under arrest by the U.S. consul. But then, surprisingly, the captain was also arrested, and charged with murder. At the trial in New York, Ludlow held that he had quelled a true mutiny. In this he was fervently supported by his fellow masters. "If Captain Ludlow is convicted, I will never again set foot on a whale ship," declared a master. But with only four mutineers out of a score of crewmen, it was difficult to prove that any attempt had been made to seize the ship.

The judge's decision was truly Solomonic. Captain Ludlow was exonerated as acting within his rights, and the three surviving seamen were found guilty of assault rather than mutiny. But to whaleship captains, the all-important thing was that the rule of the master at sea had been upheld.

When the brief but bloody melee began, the *Triton*'s third mate, Elihu Brightman, was asleep, unnoticed, in the waist whaleboat. According to one account, Brightman, armed with a lance, waited to confront the renegade Manuel. When the chance came, the mate made one thrust with his razor-sharp lance and pinned the man to the deck. The sight of their impaled leader demoralized the islanders, and the whalemen drove them over the side. The Portuguese outlaw was fed to the sharks.

Assuming that the captain and his boatmen were dead, Brightman and the crew took the *Triton* to Tahiti. The *Alabama* out of Nantucket later arrived at Sydenham Island and rescued the missing men. The islanders, no longer incited by the evil Manuel, had spared their captives' lives.

Two years later the *Charles W. Morgan* was drifting in a calm about five miles off the same island and the second mate, hanging in the lookout hoops with Nelson Haley, was recounting the story of the *Triton*. Suddenly he broke off the story — "Damn them, here they come!"

And come they did — at least 500 hostile "blue-skinned savages," as Haley remembered them, in some 50 war canoes. On deck, Captain Samson supervised the *Morgan*'s defense. At his command, crewmen slashed with cutting spades, repulsing natives who tried to climb aboard by grabbing the chain plates. The canoes backed off and surrounded the ship, evidently content to wait until a current took the *Morgan* aground on a reef encircling the island. Meanwhile the natives screamed and made insulting gestures; in one canoe a portly islander arose, shouted a local obscenity, turned his back, placed his hands on his hips and bowed low to present his broad behind. As Haley described it, "no clothes obstructed the shining mark." Captain Samson raised a shotgun loaded with bird shot, took careful aim and fired both barrels. The islander arched gracefully into the water.

The fallen warrior's paddlers beat a rapid retreat. But now the men in the *Morgan*'s bow could see bottom as the current carried the whaler toward the reef. The islanders sensed victory, and half a dozen canoes darted from the pack and came for the *Morgan*'s sides. Again the whalemen beat off the attackers with cutting spades. The ship nudged over a rim of coral as the islanders shouted in unison, triumphantly whirling their paddles. And then, at the last possible instant before going aground, the *Morgan* slowly began to move. The current, diverted by the edge of the reef, had turned and was taking the ship with it. Soon the *Morgan* was in 15 feet of clear water and the bottom dropped out of sight.

"When we saw all danger past," Haley recalled, "did we not yell in derision to those blue-bellied beggars, who had stopped their clatter on seeing the ship pass what they made sure would be her doom!"

However narrow the escape had been, and however deadly the possibilities, Nelson Haley, at least in retrospect, thought of the *Morgan*'s triumph over the contemptible savages as something of a lark. It was a luxury of attitude that Haley, young, unmarried and filled with adventuresome juices, could well afford. But for others among the whalemen, life in Pacific waters was a much more sober affair, with an added burden of responsibility — for by the mid-19th Century, in ever-increasing numbers, the wives and even the children of many whaling masters had come to share both the delights and the dangers of "the other side."

A whale of a show for stay-at-homes

New England was intensely proud of its role as whaling capital of the world. Not one in a thousand Yankees ever saw a whale, but in the mid-1800s the stay-at-homes were given an opportunity to experience some of the thrill of the chase and the wonder of the whalemen's exotic haunts. All they had to do was wait for Purrington & Russell's great panorama to hit town. And then for 25 cents they could take *A Whaling Voyage round the World*.

Ancestor of the moving picture, the panorama was a 19th Century art form by which historic events and faraway places were depicted on huge rolls of canvas that were hand-cranked, scene by scene, across a theater stage while a narrator described the action. *Burning Moscow* and *The Battle of Gettysburg* were famous mid-century panoramas. But the masterpiece—more than eight feet high and 1,295 feet long—was Purrington & Russell's whaling opus.

Self-taught artist Benjamin Russell had embarked on the New Bedford whaler *Kutusoff* in 1841, and for three years had sketched the whaling saga. Back home, he enlisted house painter Caleb Purrington to put the epic on canvas.

It was, as scenes on these pages indicate, a whale of a show. "It is not too much to say," historian Samuel Eliot Morison later commented, "that it is a pictorial counterpart to Herman Melville's classic *Moby Dick*."

The opening scene in the quarter-mile-long panorama of a whaling cruise, this scroll displays the waterfront of New Bedford, capital of the industry. Though a few liberties were taken with later landscapes, every house and steeple here had to be rendered exactly, or New Bedforders would have scorned the show.

Encountering right whales in the North Pacific, the Yankee fleet is soon engaged in furious action. As the painting was slowly unwound, audiences viewed a whale being stripped of its blubber (left), a crew hoisting a whale's head to remove the baleen, and another whale rearing and about to swamp a whaleboat.

Stopping at the Azores in the North Atlantic, Yankee whalers stocked up on fresh fruit, vegetables and water. The Western Islands, as these Portuguese possessions were called, were a good place to get rid of greenhorns who could not take the rough life, and to sign on skilled Portuguese mariners in their stead.

Spouting like a giant sperm whale, a volcano named Cano Peak in the Cape Verde Islands erupts, casting its eerie orange light onto an assortment of square-rigged vessels and lateen-rigged fishing boats. As narrators pointed out to audiences of the panorama, Cano—rising some 9,000 feet above the sea—was twice as high as Naples' famed Vesuvius.

104

Off the bleak spire of Cape Horn (left), a Yankee whaler, her sails blown away, goes down by the head in heavy seas typical of the region, while a ship stands by to rescue the crew. The narrator could be counted on to advise the audience in the most dramatic terms that immersion for more than a few minutes in the always frigid waters meant certain death.

Más a Tierra (right) in the Juan Fernández Islands, 470 miles off Chile, was a favorite watering place for whalers on Pacific voyages. Panorama narrators had plenty to say about the spot: it was the site of a famous marooning that in 1719 led Daniel Defoe to write Robinson Crusoe.

Kealakekua Bay in the Sandwich Islands presents an idyllic scene as a whaleship lies at anchor surrounded by dugouts filled with islanders eager to barter.

But less than a century before, England's great navigator, Captain James Cook, had been savagely murdered on this very beach.

Nuku Hiva in the Marquesas appears as a thriving port with sturdy buildings and a well-drilled French garrison to maintain order. Artist Benjamin Russell's path may have crossed that of Herman Melville here. Melville jumped ship at Nuku Hiva in 1842, and later wrote about his adventures with cannibals and island maidens in his novel Typee.

Arriving safely home in New Bedford with a full load of oil and baleen, a Yankee whaler is being towed into harbor by the proud new paddle-wheeler Massachusetts, churning along under wood-burning engines. At the time the panorama was painted, New Bedford was home to some 250 whalers; scarcely a day would pass without a ship arriving or departing.

The whaling wife: life alone or life aboard

top many of the captains' mansions rimming the harbors of New England whaling towns stood a structure that was sign and symbol of a haunting loneliness. It was a rectangular, railed platform, reached by a ship's ladder through a trap door in the roof. There, for hours at a time, when her husband's ship was expected home, the wife of a whaling master would stand solitary vigil, gazing out to sea for sight of the familiar shape of the vessel returning safely from its long voyage. Because some loyal women kept watch for ships that were long overdue—and that often proved to have gone to the bottom of remote seas—the platform was later known in some seaport towns as a widow's walk.

The wife of any seafaring man had to resign herself to long periods of loneliness—and to the very real possibility that she might never see her husband again. A merchant seaman's wife might not lay eyes on her husband for a year or more; a navy wife might suffer through painful months while her husband was on extended patrol or on foreign station. But these periods of separation were brief compared with those endured by the wives of New England whalemen.

A whaleman's wife could routinely count on at least three, and sometimes four or five, melancholy years of waiting and wondering while her husband pursued his intensely dangerous calling in the Pacific, 10,000 miles from home. Lydia Gardner was married to Captain George Gardner for 37 years; during that time he spent less than five years at home in Nantucket. Another whaling captain's wife calculated that during 11 years of marriage she had seen her husband for less than a year. One whaling yarn involved a skipper who was about to embark on a short North Atlantic cruise. When reminded that he had forgotten to kiss his wife goodbye, the old salt replied, "What's ailing her? I'm only going to be gone six months."

The pain of absence was aggravated by the difficulty of communication. Correspondence between wife and husband moved slowly down the Atlantic, around the Horn and across the Pacific, and often failed to reach its destination at all. One New England wife sent off more than 100 letters to her whaling husband during a three-year voyage; he received only six of them.

In the mid-1850s, when Honolulu became an important whaling town with many of the amenities of home, including a post office, this dismal

From a railed perch atop their home, two women watch for the familiar square-rigged sail announcing the return of a whaling husband, son or brother. While these platforms, later called widow's walks, were designed primarily for observation, they also provided access to roofs for fighting chimney fires. Many Nantucket home owners kept buckets of sand or water at the base of the ladder leading up to the walk for this purpose.

record improved greatly. But in the early years the whalemen used mail drops on various uninhabited Pacific islands. One popular drop was on Charles Island in the Galápagos; it consisted of a large box covered with the shell of a giant tortoise and nailed to the top of a post at the head of a sheltered cove; whalemen called the place Post Office Bay. Outward-bound whalers would stop at Post Office Bay to leave packets of letters from New England for vessels already in the Pacific. Ships whaling the area would stop to pick up these letters and to drop off others that would be collected and carried home by whalers whose cruises were ending.

Even when letters were delivered they did not always solve the problems of distance and separation; they could bring more anxiety than comfort. A letter from home might report the illness of a loved one, leaving a father or husband deeply troubled—and with the agonizing knowledge that it would be months before he got further news. When Captain Presbury Luce of the *Emily* received a large packet of letters in Talcahuano, Chile, a paralyzing dread of getting bad news along with the good kept him from opening them for 20 days. When he did, he learned that his wife had died, just before her 35th year. They had been married 12 years and had been together only half of that time.

"Home sweet home and those we left behind us are constantly on our minds," one whaling captain wrote in his log in a rare emotional outburst, "little do those on shore know a sailor's feelings, separated from all that they hold dear on earth, with almost a certainty of being apart for three or four long years—enough to make a man's hair grow grey at the thought of it."

The unnatural existence was hard on husbands, even harder on women. Infant mortality was high in the 19th Century, and many a child was born, lived for a year or two and died before his father ever saw him. And there was the terror of receiving a letter that had been addressed by a husband's shipmate. In one such letter, Eliza Inott of Nantucket was told of the death of her husband, killed by a whale, when their son was three years old. She never remarried—but during the rest of her life the perils of whaling continued to claim men from her family: her little boy grew up to die of yellow fever on a whaler off Honduras, and two of her grandsons were also killed whaling, one when his ship was lost in a storm, the other of unrecorded causes.

Whaling deaths and voyages turned some New England coastal towns into largely female communities. At the height of the whaling industry, women actually outnumbered men by four to one on Nantucket. Many found more productive ways to fill their lonely hours than treading the widow's walk. Women had no say in the island's government, since they did not have the vote, but they ran much of Nantucket's commerce. The wife of a captain would keep his books and market the exotic goods he brought home from the Orient and other faraway places. Nantucket's groceries, bakeries, dry-goods stores, apothecaries and other shops were, more often than not, run by women. The main thoroughfare, Centre Street, was known as Petticoat Row for all the stores along it that were managed by women—who simultaneously maintained homes and reared children.

But for some wives this was not enough. They were unwilling to stay

William and Rhoda Cushman, aged five and three, the handsome children of New Bedford Captain Benjamin Cushman, stare solemnly ahead in this portrait painted after they died of scarlet fever in 1840—while their father was at sea.

in New England, minding a shop or pining alone, while their husbands sailed off for three or four years at a time. These wives chose a more dramatic solution to the loneliness of a whaling marriage: they went to sea with their men. By mid-century, nearly 100 New England wives were in the Pacific, and women who had never been to New York or Boston found themselves in such odd and exotic places as Rarotonga and Lahaina, Talcahuano and Aitutaki, Ponape and Okinawa (then known to whalemen as Great Loo Choo).

The odysseys of these seagoing wives provide some of the most fascinating yet least known stories of the great era of whaling. In their cramped quarters they gave birth and raised families. Although some found it difficult to win acceptance on board ship, they came to play a valued role. They suffered everyday discomforts and survived storms. They encountered strange peoples with strange customs, and they created little patches of New England in faraway places. Many of them kept journals—personal, idiosyncratic records that, unlike the cryptic logs of their husbands, often offered refreshing insights into whaleship life.

Nantucket's Mary Hayden Russell was the first woman to follow her man to sea, and she probably had more reason than most: her husband, Laban Russell, was one of the many Yankee captains whose home ports were on the other side of the Atlantic. For a time Mary lived in Milford Haven in south Wales in order to be with her husband between voyages. But his sojourns were so infrequent that in 1817 she boarded his ship, the *Hydra*, to accompany him—and her 12-year-old son, William, whom his father had signed on as cabin boy—during a three-year cruise to the Pacific. She seems to have kept no journal on that voyage, but the experiment was apparently enough of a success for her to join her husband and William, by now a 17-year-old harpooneer, in 1823 when they set forth in a second vessel, the *Emily*, for the Pacific. This time she took along her seven-year-old son, Charles, to share in the family experiences—and she faithfully recorded all of them in a journal.

Mary was by no means a complaining sort, and she never let on how cramped she must have found her abode. A captain's quarters usually consisted of two small rooms, which together were smaller than the parlor in a modest New England home. There was a cabin that served as both an office and a sitting room, and an adjoining stateroom with just enough space for a three-quarter bed for husband and wife to share. Next to these two rooms—or sometimes taking up a corner of the stateroom— was a luxury that only captains enjoyed: a tiny privy.

In the Russells' case, a shelf above the traditional horsehair sofa in the main cabin was converted into a narrow berth for little Charles. The cabin's only outlooks were the small stern windows and an overhead skylight with panes of glass set in a barred frame. For most of the voyage, Mary Russell would see through these apertures nothing but patches of rolling water to the stern and, above, the steady parabola described by the mizzenmast and its rigging.

Simply unpacking was a major problem for whaleship wives. One trunk might remain in the quarters to serve as a table or bench. The contents of others were somehow crammed into two or three drawers beneath the bed before the trunks themselves were stowed in the ship's hold. Mary's journal does not relate what reminders she brought from home to make her new abode more livable. But she probably brought something; other wives did: a geranium or two, a rocking chair, a sewing machine, a portable parlor organ on which to lead the family and perhaps the officers in Sabbath hymns.

Mary's life aboard the *Emily* was one of sharp contrasts, of long periods of tranquillity shattered by moments of exhilaration and crisis. The first time a whale was sighted, Mary Russell wrote, she was so alarmed by all the shouting and pounding that she thought the ship was sinking.

Another day she wrote: "The evening proving fine, I had a chair placed on deck to see the sun set. My whole mind was engaged in contemplating the magnificence of the sun when I heard a scream from my dear little Charles, who had the minute before left my side. Before I had time to inquire the cause his brother brought him to me with his arm broken just above the wrist joint. Such an accident on the land would have been distressing, but what were my feelings when I saw the child writhing in agony and no surgeon on board. He had been to the cabin and as he was returning a sudden lurch of the ship caused him to fall with his weight upon his arm, which snapt it. His dear father, with that fortitude and presence of mind that seldom forsakes him, took him below and, with a man to steady the arm, set it and splintered it up."

Setting broken bones was commonplace to the masters of whaling ships, which carried no ship's doctors. Internal ailments offered more formidable problems, and treatment was often reckoned by guess and by God. In such ministrations Mary Russell could help and, like many another whaleship wife, she found immense satisfaction in her useful role. "I have often had reason, since I left Nantucket," she wrote, "to bless the little knowledge I had of medicine, as it has contributed to take a great care off the mind of my husband. He examines the cases and reports them to me; this is his part and I am happy to say that the medicines I have administer'd have never failed of their desired effect."

Through these New England women ran a strong streak of faith, and it sustained them in their vicissitudes. During one gale, while the sea raged "with indescribable violence," young William Russell visited the cabin to comfort his mother—even while warning her that their lives were in danger. "The hull of the ship seems not to have sustained any injury," she recalled that he said calmly, "but should it be otherwise, and this night be our last, we will go trusting in the mercy of God." When William Russell returned to the desperate work of his ship, he left his mother confiding her fears and finding strength in prayer. Again, upon watching Laban and William Russell lower their boat and go out to attack whales—"my husband, my son expos'd to these monsters of the deep"—Mary suffered agonies of anxiety. Yet she could still find a certain solace: "What a comfort at that moment to reflect that they were in the hands of God who was as willing as able to protect them. I could truly say: 'They that go down to the sea on ships, that do business in the mighty waters, they seeth wonderful works of the Lord.'"

Mary's voyage ended when the *Emily* returned to London in 1825. By that time Mary had become very much at home aboard a whaler—and

After eight years of reading love poems composed by her faraway husband, Desire Fisher joined him in 1858 aboard the Navigator for a Pacific whaling voyage. She returned to Edgartown, Massachusetts, after a year: a mutiny convinced her that solitude was preferable to life at sea.

Cheerful Caroline Mayhew, who spent 15 years on whaleships with her husband, William, made a hit with South Sea islanders by fashioning calico pinafores for their children. They in turn showered her with gifts—including a baby kangaroo.

The whaler Niger had taken "upwards of 100 sperm," wrote Charlotte Jernegan from the Pacific to a friend back home in 1857. She was so enthusiastic that she only mentioned as an afterthought, "I have a little son, five weeks old today."

"The captain's wife is not so much frightened as the hands," wrote Captain Shubael Norton of his Susan's bravery in a hurricane in 1858. Susan Norton's fortitude was borne of her trust that the experience would make her "a better woman."

she had formed her own philosophy, sharpened by her perceptions of the perilous profession of her husband and son. On November 18, 1824, she had written a letter to her daughter describing a fearful afternoon during which she watched her son's boat being dragged across the water by a whale, then saw her husband's boat disappear in a thunderstorm and remain out of sight for four hours.

"Think, my dear Mary Ann," wrote Mary Russell, "how anxious I must have been, and how happy I was to see your dear Father once more. He had not a dry thread in his clothes. I thought: 'this is the way these sons of the ocean earn their money—that is so thoughtlessly spent at home.'

"Could some of the ladies whose husbands are occupied in this dangerous business have been here these few hours past, I think it would be a lesson they would not forget. It would teach them prudence and economy more powerfully than all the books ever written on the subject since the invention of printing."

In at least one sense, Mary Russell was fortunate: she had a strong stomach and was rarely afflicted by a humbling human ailment for which neither the medicine chest nor the Bible held a palliative. "My poor Emma," wrote Captain Horatio N. Gray of the whaling bark *Cossack*, "is suffering much from her old enemy sea-sickness and there seems to be no cure for her. Whiskey gives some relief, as does Brandy and water—but it is only temporary. I am very sorry that I ever gave my consent to her coming on this voyage."

Lucy Smith, wife of Captain George Smith of the *Nautilus*, was still unpacking her things when three-and-a-half-year-old Freddie announced that he didn't feel well. "I asked him where he felt sick," she recalled, "and he rubbed his stomach, head, feet, he did not know just where, he felt badly all over. He soon threw up and then lay down and went to sleep." Freddie, it developed, recovered from his seasickness more easily than his mother, who was stricken shortly afterward. She was so sick, she wrote, that "I could not raise my head. My husband carried me on deck wrapped in a blanket to a little room fitted for me where I got fresh air and after a few days' airing I was all right."

On her first day at sea, Eliza Williams, wife of Captain Thomas Williams of the New Bedford whaler *Florida,* wrote plaintively in her journal: "I think I am getting Sea sick." Perhaps she was—but the fact that she was five months pregnant may also have had something to do with her queasiness. A pretty little woman who weighed less than 100 pounds and could stand under the outstretched arm of her six-foot-three-inch husband, Eliza had already borne him two sons during his whaling absences. Thomas Williams did not see his first son until the boy was two years old. So much did Eliza miss her husband that she asked him to refrain from writing sentimental letters—they were more than her lonely spirit could stand. And when Williams sailed on his third voyage as a whaling captain, Eliza was determined to accompany him, even though it meant leaving sons Stancel and Henry at home with their aunts.

"In company with my Husband," Eliza started her journal, "I stepped on board the Pilot Boat about 9 o'clock the morning of the 7th of September, 1858, to proceed to the Ship *Florida,* that will take us out to Sea far from Friends and home, for a long time to come."

Eliza had barely started to recover from her indisposition when, on the third day out at sea, foul weather again laid her low. "I call it a gale," she wrote, "but my Husband laughs at me, and tells me that I have not seen a gale yet."

She soon would. Her husband had a reputation as a lucky captain, and even before reaching the Pacific the *Florida* began taking whales. On November 9 the ship was surrounded by so many whales it seemed to Eliza that "there were more than a hundred." The boats were lowered, and one shortly returned with a whale and a smashed bow. When darkness came, some of the boats were still out over the horizon and Eliza "was fearful of their safety." To signal the ship's whereabouts, she reported, the men "built a fire in the try works of bits of tarred rope and scraps, which made a nice blaze. They halloaed to them and got the horn and blew. My Husband told me that he was not alarmed about them for it was calm and they could keep track of the Ship, as she would not go out of the way." Soon the boats did return, two of them with a whale. "I was glad when they came," Eliza wrote, adding: "My Husband tells me that I will see worse times than these before I go home."

There were five whales to render, and the cutting in started at dawn, after the exhausted whalemen caught some sleep. Then, just as the work got under way, a gale came up—and this time there could be no doubt about the force of the storm. Bracing herself against the wind, Eliza Williams stood on her little section of the afterdeck and watched her men—for so she had come to consider them all—struggling to save their catch. Her description of the scene was all the more graphic because the witness was impressionable:

"I thought it was impossible for them to work at all, with the waves dashing against the Ship and those huge monsters moving up and down in the water, sometimes so covered that you could scarcely see them. But they worked on and did not cease. There was a complete din of noises on deck—the wind, the rain, the Officers shouting to the Men, the mincing machine, and altogether it was a confused place. One of the stagings gave way and pitched the first Mate into the water, but he did not seem to mind that, for he was up and to work again, wet as he was; in fact they were all as wet as they could well be. The work went on through the stormy day and night, and by next morning the weather had improved."

That afternoon Eliza was treated to a visit to an unfamiliar part of the ship, the blubber room. "The Mate came to me," Eliza recorded, "and wanted me to go with him and take a look down in the 'reception room,' as he termed it. I went, and I could not refrain from laughter, such a comical sight! There the Men were at work up to their waists in blubber. The warm weather had tried out the oil a good deal and made it soft. I don't see how they could stand in among it, but they were laughing and having a good deal of fun."

Probably out of concern for his wife's pregnancy, Captain Williams had chosen to take a longer and slower but more tranquil route than usual—around the Cape of Good Hope instead of around the storm-swept Horn. Although her journal was silent on the subject of her pregnancy—she was reticent about discussing it, even in her diary—Eliza's time was approaching. In December she started omitting days in her entries; she made no mention at all of Christmas. Then in an undated January entry, written after the *Florida* had reached Mangonui, New Zealand, she observed, "It is now about a month since I have written anything in my Journal and many things have transpired since then."

In summarizing those neglected events, she told of encountering and speaking another ship and of experiencing a terrible storm. Then, almost casually, she wrote: "We have a fine healthy Boy, born on the 12th, five days before we got into Port." She said not a word about her oncoming labor at the height of the storm, or about the long and painful delivery in the cramped cabin, with only her husband to act as midwife.

The birth of her son, whom she named William, left Eliza so weak that "I had to stay on board the Ship for two weeks." The Mangonui harbor master sent his wife aboard the *Florida* every day to care for the baby, and Eliza gratefully noted, "She did everything she could for me till I was able to go to her house." By February 5, with her baby less than a month old, Eliza was back on board the *Florida*, headed into the Pacific, and writing, "It has been quite calm and I have been walking on deck a little while. Our Boy is well and grows finely."

During the next two years Eliza Williams made it clear that, despite the hardships, she was savoring an experience granted to few women of her —or any other—generation. Once, when the *Florida*'s men were cutting in a whale, she wrote delightedly: "My Husband wanted me to walk into the whale's mouth. He pushed me in a little way, so I think I can say that I have been inside of a whale's mouth. Six or eight people could go inside," she marveled, "and sit down at one time."

Wherever whales could be found, so went the *Florida*—to the frigid Okhotsk Sea off Siberia in summer, to the central and southern Pacific in winter, along the coast of South America in spring and autumn. Eliza was intrigued by the oddity of snow and ice in the Okhotsk Sea in late spring and summer. On one occasion a group of ships anchored in a harbor and a number of the captains went ashore. "My Husband and some of the others," she wrote, "were snowballing, having some fun." The Okhotsk shoreline was "rugged country and awfully grand," and Eliza had the rare privilege of taking a walk on a beach. "I found some very pretty wildflowers and some nice mosses," she reported. "It was delightful roaming around on shore a little while after being aboard of the Ship so long." And, she added, "I sat down on the skull of a whale."

In the Carolines she and her husband visited the native king of Strong Island. "He has hung all about the walls of the house, war clubs, hatchets, and curiosities that he has gathered together," she wrote, "and overhead fancy Canoes all decorated. They look quite pretty. It made me think of a museum."

By the time the *Florida* reached Japan in May 1859, six years after Commodore Matthew Perry's historic visit and 14 years after the first whaler called at the islands (pages 82-83), the Japanese were becoming quite friendly to foreigners. Eliza was fascinated by some officials who boarded the ship at Hakodate. "They were dressed quite nicely though quite singularly, to me,'' she noted. ''Their dress is quite loose and slouching, very loose pants if they can be called such, and a kind of loose cloak with very large sleeves.'' She marveled at the visitors' wooden clogs, which made ''a great deal of noise'' on the *Florida*'s deck, and at their Japanese swords and the long, sharp dirk knives they carried in sheaths on their belts. And she happily reported that they ''were highly pleased with the Baby. They crowded around him, feeling of him, and talking and laughing with him.''

''Willie,'' as she called him, was a delight to her as he grew from infant to toddler. When he was two years old, Eliza wrote: ''He almost lives on deck, and he is generally full of mischief when he is up there, throwing his Shoes and Cap, or something, overboard.'' His father took him on a duck-hunting expedition on the Pacific coast of Mexico and his mother proudly recounted that ''Willie seemed as fresh as when he left in the morning, though he had been gone all day. His little shoes were wet and full of sand, so that he limped when he walked, but he did not seem to mind it. All he could say was 'Papa, bang go Ducks!' ''

Yet for all those memorable family moments, there were times when Eliza knew a gnawing homesickness for America. One November she confided to her journal: ''Thanksgiving, that day of all others that we take so much comfort in at home with Friends, is over now; we knew nothing about it here.'' Of Independence Day she lamented: ''Oh how I would like to be at home and enjoy this day with family and friends. We cannot celebrate it here with any degree of pleasure.'' And on another special occasion she fairly cried out in her longing for her sons at home: ''Today is Stancel's birthday. Oh, how I wish I could see him! Words are too feeble to express the great desire of my heart to once more set eyes on those Dear Children, that Dear Home, Parents and Friends.''

Eliza Williams was oppressed, too, by the constant awareness that men for whom she had come to care might not return from a day's work in the whaleboats. One morning, after lowering the *Florida*'s boats, Captain Williams came below to tell Eliza that the ship's best harpooneer had been caught in the line and dragged beneath the surface by a whale. ''He has taken more Whales for us than any other Man aboard of the Ship, and never missed one,'' Eliza wrote. ''But it is not his services alone that I think of; it was such an awful death to die. He was a colored Man. He was a very pleasant Man. I never went on deck but what he had a smile on his face.''

Eliza's unending dread, of course, was that a similar misfortune would befall her husband. And one summer night she thought it finally had. Until well past 2 a.m. Eliza walked the deck of the *Florida* while a gale screamed around her, waiting and watching for some sign of Captain Williams' boat, which had disappeared into the storm while its crew was pursuing a whale. As the ship pitched and rolled, seas poured over the bow, washing the afterdeck, where Eliza stood her watch. ''I could not think what had become of them,'' she wrote the next day. But the entry had a happy ending. The boat had been stranded in an estuary by a receding tide, and Captain Williams had had to wait to be floated again. He finally made his way back to his ship and his worried wife at 3 a.m.

By March 26, 1861, the *Florida* had been at sea almost continuously

Only up to a point was Phebe Ann Pease willing to cruise with her husband, Captain Henry A. Pease, on the Cambria. After three years she left the ship in 1861 and took passage from the Pacific back to Martha's Vineyard so that their daughter, Grace, might be born on New England soil.

A painting of the Cambria, done by a Chinese artist in the Orient, held a place of honor in the Edgartown parlor of Phebe Ann Pease as a proud reminder of the time she had spent on board the vessel with her husband. So high a value did Mrs. Pease place on the picture and other mementos of her whaling days that, according to her niece, the wood stove in that "chill, hushed room" was never fired lest ashes fall on her cherished collection.

for two and a half years. On that date, without any previous hint of the impending event, Eliza's diary announced: "We have an addition to the *Florida*'s crew in the form of a little Daughter, born on the 27th of February in Banderas Bay on the coast of Mexico." Eliza continued to sail with her husband, always taking one or another of the children, for 13 more years, until at last in 1874 she decided that she had seen enough of whaling and shipboard life. From then on she remained at home in Oakland, California, to which the family had moved in 1866.

An Eliza Williams might win acceptance in the male world of whaling by her cheerful and caring presence. However, many whalemen derisively called the ships carrying captains' wives "hen frigates," and most women had to earn the regard of the other officers and crewmen. Some never did. Jane Worth of the New Bedford whaler *Gazelle*, for example, irritated a fourth mate who disliked whaleship wives in general and Jane in particular. "The whistle of a gale of wind through the rigging," he wrote in his diary, "is much more musical than the sound of her voice." She was childless and evidently lonely; he once caught a glimpse of her that left him less moved than disdainful. "Looking into the after cabin today," he wrote, "I saw in a cradle two dolls and beside them sat a pretended mother singing and talking to them as a little girl would. It is useless to write any more about that but I have formed my opinion.''

Jane Worth scored a temporary triumph one day in the Banda Sea after she had somehow come into possession of a familiar New England delicacy. With the temperature ''about 100 in the shade,'' the mate wrote, ''time hangs heavy. I had a taste of rhubarb pie today at dinner time. Mrs. Worth reckoned I had better eat something that would taste like home. Found it very palatable, wish we could always have something that would taste like home.''

But the opportunity to extend such heart-winning gestures was infrequent: no matter how much the men may have enjoyed home cooking, ships' cooks generally resented the presence of wives in their galleys. And soon the fourth mate was back at his faultfinding. ''The best place for a female is at home,'' he wrote. ''Who would not sooner be a bachelor than put up with the like?''

More successful was Mrs. Charles Grant, who sailed with her husband for 32 years and on one occasion earned the crew's gratitude by spotting a whale while she was hanging out her wash. Before the sewing machine of another whaling wife, Lucy Smith, finally rusted out in the salty, humid air of the Pacific, she had made a ship's flag for the *Nautilus*, a jib for one of the boats and a mainsail for another, a canvas cover for the chronometer box, a covering and cushions for the cabin chairs, and a tablecloth for the dining cabin. Another woman who won the gratitude of a ship's company was the *Powhatan*'s Caroline Mayhew, a doctor's daughter. When smallpox broke out on board her husband's ship, she refused to be put ashore on a nearby island, remaining to tend the sick. As the disease spread, Caroline's husband, William, was among those stricken. Caroline nursed him back to health—meanwhile taking over the ship's navigation. Finally she too was taken ill. When she recovered, crewmen of the *Powhatan*, saying that she had saved their lives, showered her with gifts of scrimshaw in thanks.

Charity Norton, a lady of formidable appearance but warm heart, rounded the Horn six times and was beloved by the whaleship *Ionia*'s crewmen—if only because she courageously stood between them and the tyrannies of her husband, John. Like many harsh masters, Captain Norton was plagued by desertions, and at one South American port 14 of the *Ionia*'s crew jumped ship. Captain Norton sailed without them, then returned to port. As he had guessed, his deserters had come out of hiding during his brief absence; 12 of them were caught and brought back on board the *Ionia*. Once more at sea, Captain Norton had the men bound to the rigging. Charity Norton came on deck, looked at the scene and asked, ''John, what are those men in the rigging for?''

''I'm going to lick 'em,'' said John.

''Oh no you're not,'' announced Charity firmly—and in the face of her determined opposition the terrible-tempered Captain Norton backed down. As far as anyone knows, the men were never punished.

For children a whaleship was a wondrous place to grow up. They marveled at the strange-looking fish that were caught and lay thumping on the deck, some changing colors when they came out of the water. The

Laura Jernegan—whose portrait graces this clock owned by her father, Captain Jared Jernegan—spent hours on board the Roman penciling accounts of life at sea. In the epistle at right, dated March 6, 1870, the well-traveled tot tells her grandmother that the family feasted on fruits with a Pacific island queen and had a "nice time" visiting the governor of another island.

MARCH 6. 1870.

MY DEAR GRANMAR.
WE ARE AT SEA NOW. I EXPECT WE
SHAL BE AT HONOLULU IN ONE WEEK.
I HAVE A LITTLE KITTEN. SHE IS
GOOD. SHE IS BLACK AND WHITE.
I CAN WRITE A LITTLE. I AM GOING.
TO HAVE A TEASET. PRESCOTT IS
OUT ON DECK. WHERE ARE YOU NOW.
I SHOULD LIKE TO KNOW. WHERE.
I SUPPOSE YOU ARE IN EDGARTOWN.
I AM GOING TO WRITE AUNT EVA A
LETTER. YOU HAVE HAD TO WAIT.
A LONG TIME FOR YOUR
LETTER. WE WENT TO AN ISLAND
NAMED OHITAHOO AND STAID
EIGHT DAYS. WE WENT TO THE
QUEENS PALACE. AND SHE MADE
A FEAST FOR US. MAMA WAS
THE FIRST WHITE WOMAN THAT
EVER WAS ON THE ISLAND. WE
HAD TEN DIFFERENT KINDS OF
FRUITES. I WILL NAME THEM.

ORAGES. BANANAAS. PINEAPPLES.
PLANTAINS. BREADFRUIT. COCONUTS.
LEMONS. MUMMIE APPLES. LIMES.
GUAVAS. WE WENT TO ANOTHER
ISLAND NAMED NOUKAHIVA. AND
PRESCOTT. AND I HAD A NICE TIME.
THE WHITE PEOPLE HERE
ARE ALL FRENCH. WE WENT TO.
SEE THE GOVERNOR. AND THE
QUEEN. AND THE SISTERS OF
CHARITY. PAPA HAS TAKEN 4
SPERM WHALES THAT MADE
83 BARRELS. I HAVE HAD A
PRESENT OF A VERY HANDSOME
WRITING DESK. AND A SILVER FRUIT
KNIFE. I HOPE I SHALL HAVE A
LETTER FROM YOU. AND AUNT EVA.
I SEND MY LOVE TO YOU AND
AUNT BINNIE. AND ALL THE REST.
PRESCOTT AND I SEND YOU
LOTS OF KISSES.
FROM YOUR DEAR LAURA.

rigging was an ever-present temptation to every boy, and few mothers were able to keep their sons from scrambling up the ratlines, though a brief stay in the rolling crosstrees usually sent the younger ones scurrying back to the deck. Aboard the ships there seemed an almost infinite number of hidden places to explore—when and if permitted. Captain Thomas Mellen of the *Europa* drew a chalk mark across the afterdeck and forbade his children to cross it lest they get in the way of the crew. Laura Mellen, aged six, obeyed. But her younger brother, Archie, slipped across the line whenever his parents were not watching.

In addition to his own family, Captain Mellen also shipped as a cabin boy the 12-year-old son of the *Europa*'s first mate. Jamie Earle turned out to be a prankster who was totally unimpressed by the lofty position of his benefactor. He heated the grip of the poker the captain used to light his pipe, and he sprinkled tacks on the hatch cover, which was the captain's favorite seat. Crewmen remembered with evident pleasure the spectacle of Captain Mellen chasing Jamie around the deck—with tacks still sticking to the white duck seat of his trousers.

When the whaleship *Roman* sailed from New Bedford in October of 1868, it carried Captain Jared Jernegan's wife, Helen, and, in tiny bunks protected by latticework to keep them from tumbling out, the couple's six-year-old daughter Laura and two-year-old son Prescott. For the children the voyage became a kaleidoscopic delight—with one interlude of stark terror. They spent long, pleasurable hours perched in the ship's deep stern windows, watching the sharks in the vessel's wake. At the Juan Fernández Islands they peered into Robinson Crusoe's cave. By the time they reached Honolulu, Prescott was so accustomed to being on board ship that he was afraid of land; he cried when asked to step ashore from the ship's boat. At one of the Pacific islands naked, tattooed islanders came aboard and presented Prescott with a black piglet, which was soon following him all over the ship.

Laura was tutored by her mother, a former teacher, and kept an almost daily journal that gave a child's-eye view of life aboard a whaler:

Friday, February 10, 1871. "it is quite rough to day. But is a fair wind. We have 135 barrels of oil, 60 of humpback and 75 of sperm. We had too birds, there is one now. One died. There names were Dick and Lulu. Dick died. Lulu is going to. Prescott has got a little dog, its name is Tony. We have not seen a ship since we left Honolulu. Prescott is playing with Papa. I am in the forth reder, and the fifth righting book. Good Bye For To Day."

Saturday, February 11, 1871. "Lulu died last night. It is quite smooth to day. It is most dinner time and I am very hungry. We are to have fresh mutton for dinner. Papa put up a hammock for Prescott and me. Mama is going to make a sack for herself. Papa is fixing the sink. Good Bye For To Day."

Sunday, February 12, 1871. "it is Sunday. it rained last night. Papa made a trap and caught 5 mice, and mama has some hens that laid 37 eggs. Good Bye For To Day."

Dick and Lulu, hens and hammocks—and one terrifying occasion that Laura could not bring herself to record: mutiny. While ashore on one of the islands, 16 of the *Roman*'s men got drunk, returned to the ship and tied up two of the mates on deck. Jared Jernegan loaded his rifle, ordered his wife and children into the sleeping room and waited for the attackers to swarm down the companionway. While Laura and Prescott cowered in horror, the besotted mutineers climbed down into three of the *Roman*'s boats and rowed for shore again. Captain Jernegan ordered the anchor cable slipped, and with nine remaining whalemen took the ship out to sea and safely to Honolulu.

For many Yankee whaling families, Honolulu in the mid-19th Century was a haven amid the palms and a home away from home. Throughout the golden era of whaling, some ships prolonged their profitable stays in the Pacific by putting in at the Sandwich Islands, as Hawaii was then known, transferring their oil to merchant vessels for shipment home, then returning to the whaling grounds for more hunting. And after 1848, when Captain Thomas W. Roys of Sag Harbor discovered an enormous number of bowhead whales in the Arctic, Honolulu became a natural point of departure for the new grounds.

At the start of the century, Honolulu was a settlement of grass huts and a few stores, with a population of 2,500. In the next 50 years it became a thriving little city. Between 1840 and 1860, for example, the average number of ships to ride anchor in Hawaiian harbors was 400 per year. Residents claimed, "You can walk from one end of Honolulu Harbor to the other, ship to ship, without getting your feet wet." The captains' wives soon set up a miniature New England in Honolulu, moving ashore into boardinghouses while their ships were being refitted. Others came out from home to join their husbands at this new outpost, sailing to the Isthmus of Panama and crossing to the Pacific to take ship again to San Francisco and board such vessels as the well-known steamship *Moses Taylor*, called the "Rolling Moses," for the last leg to Honolulu.

Under the intoxicating influence of the tropical islands, they indulged themselves in parties and balls that they would have disapproved of back home. There were horseback rides along the beaches, pleasant picnics, afternoons of tea and croquet, supper parties and evenings of band music to while away the time. "I kept wondering," one wife wrote, "if I had died and gone to heaven."

But they remained a salty, self-possessed group of ladies. On a visit to Honolulu, Mark Twain may or may not have been exaggerating broadly when he reported a conversation with one of the whaling wives: "I have just met an estimable lady, Mrs. Captain Jollopson, whose husband (with her assistance) commands the whaling bark *Lucretia Wilkerson* — and she said: 'While I was laying off and on before the post office, here comes

Its harbor filled with vessels and its streets lined with neat wood and coral-block buildings, Honolulu resembles a tropical version of a New England seaport in this 1854 lithograph. But there were still many things in the islands, wrote Mary Lawrence, "which put one to the blush"—such as the fact that "as I am writing, two men are close by my door without an article of clothing."

a shipkeeper around the corner three sheets in the wind and his dead-lights stove in, and I see by the way he was bulling that if he didn't sheer off and shorten sail he'd foul my larboard stuns'l-boom. I backed off fast as I could, and sung out to him to port his helm, but it warn't no use; he'd everything drawing and I had considerable sternway, and he just struck me a little abaft the beam, and down I went, head on, and skunned my elbow. I shouldn't wonder if I'd have to be hove down.' ''

When the arctic grounds opened, many captains, considering the life above the Bering Strait too rigorous for their wives and children, left them ashore at Honolulu. But not even the prospect of arctic ice and storms could daunt the more intrepid women for long, and by the late 1850s almost two dozen were journeying north with their husbands.

One of these was Mary Chipman Lawrence, who with her five-year-old daughter Minnie had accompanied Captain Samuel Lawrence when his ship *Addison* left New Bedford on November 25, 1856. She was convinced that she had made the right decision. "We are," she wrote early in her journal, "as it were, shut out from our friends in a little kingdom of our own of which Samuel is prime ruler. I should never have known what a great man he was if I had not accompanied him." She added her hope that "I may continue worthy of his love." For the next four years her confidence would not waver.

By April 18, 1857, the Lawrence family was installed in a straw house at Lahaina, on the island of Maui, with a view of crashing breakers outside their door. While Mary and Minnie sampled breadfruit and figs, the *Addison* was refitted for the Arctic, and on April 29 Captain Lawrence sailed for the Bering Strait. Aboard were wife Mary, daughter Minnie and Minnie's new dog, Pincher.

They were headed for a summer's cruise in a part of the world where at this time of year the sun scarcely set, the temperature ranged from below zero to 60° F. and one day could bring sunshine, fog, snow squalls and hail. Their quarry was the bowhead whale.

By May 8 the weather was still bitterly cold as the *Addison* approached the Gulf of Alaska, and the captain's cabin became even more crowded when the stove was set up. Through June, July and August of 1857, the *Addison*'s foghorn sounded often across the misty waters; line squalls pelted the ship with snow, sleet and hail; and a crisp, sunny day was a rare event. Mary Lawrence remained content. Recording "a succession of damp, foggy, windy, cold, dreary days, very far from pleasant on deck," she wrote that "I get along very comfortably with it, however. Have a nice little stove, a good cozy fire, a kind husband, and a dear little daughter. How ungrateful should I be to complain."

The Lawrences went north to the Arctic twice more in the ensuing three years, interspersing their summer cruises with sperm whale voyages in the Pacific during the winter months. One morning during their second voyage north, the *Addison* crashed into an ice floe. It happened during breakfast, which was at 6 a.m. Because breakfast at that hour made such a long day, Mary Lawrence usually let Minnie sleep late, making porridge for her when she woke. But this morning she rushed to rouse her daughter. "I was very calm and composed while dressing her," Mary reported, "and was ready to collect my things preparatory to leav-

Rigged with a seat and a rope sling, this barrel has been turned into a gamming chair—to lower the captain's wife into a whaleboat for a gam, or visit to another vessel. The chair was as much necessity as convenience; a petticoated lady could scarcely have climbed a rope ladder.

A whaleboat carries visitors from one ship to another in this painting of a gam between the Sea Fox, the James Allen and the Commodore Morris in the Pacific. Everyone on board a whaler found the tradition a welcome break from the tedium of endless cruising. As one whaleman described it, the news and gossip of a gam "furnish us with the materials for thought so we may not stagnate altogether."

ing the ship, as I expected we would be obliged to do." The whaleship leaked through a hole in the port bow, but the men stopped the flow with canvas and board. Three days later heavy seas opened another hole, which was also patched up—just in time for a gale the next day. "I felt anxious," Mary wrote, "not knowing how our ship would stand such a heavy sea, but she withstood it bravely; leaks no more than in a calm."

There were greater disasters, in Minnie's eyes. On their second cruise north Minnie was accompanied by a pet hen—"a hen of a very peculiar kind," Mary observed, "its feathers appearing to grow the wrong way, all in a heap." Minnie named it Frizzle. On a frigid day in May, with ice on the deck, Frizzle "departed this life. Probably the change of climate was too much for her."

An even worse tragedy had occurred on the first day of this voyage. "She lost her Frankie doll overboard," Minnie's mother wrote, "a doll that she dearly loved for its own sake and the more because it was Grandma Annie's. She cried for a long time and wrung her hands in the greatest agony." Minnie even asked to go into mourning, and her mother "was obliged to get a piece of black ribbon to tie on her arm to pacify her." Minnie was somewhat consoled, or at least distracted from her grief, by the birth of six pigs next day.

So popular had bowhead whaling become that parts of the Arctic in summer held floating communities of New England families, and a whaling custom called gamming—maybe from the obsolete word "gammon," to talk idly—flourished as never before. Long before women joined the fleet, the gam had become almost as ritualistic in its etiquette as a minuet. Passing whalers would hail each other, invitations were issued and accepted, and boats were lowered. Since all places in the boats were manned, captains remained standing, and it became a point of pride for them never to steady themselves by holding onto anything. Indeed, to show off their sea legs, most masters kept their hands in their pockets; many a captain was therefore toppled into the sea by a sudden pitch of his boat, but if dignity was lost, honor was preserved.

For the captains, the gam was at least as much business as pleasure. They exchanged information about such matters as where the latest herd of whales had been spotted, the most recent prices of sperm oil and the location of newly discovered reefs. Meanwhile, so that some semblance of watches might be maintained, roughly half of each crew joined half of the other on the deck or in the forecastle, trading their own store of information, consisting mostly of yarns—about storms and shipwrecks, about exotic islands and barren reefs, and most of all about whales.

The presence of women naturally added new dimensions to the gam. Among other things, a convenience known as a "gamming chair" was devised. Actually an armchair rigged so that it could be raised and lowered over the side of a ship, it permitted a captain's wife to board and disembark without having to climb a rope ladder and expose her ankles.

And so one morning in the Bering Sea, Mary Lawrence awoke to find herself "in a city of ships." There were 15 whalers around the Addison. Many of the captains and their wives were already well acquainted; some were neighbors or relatives from home. Just as the women had walked down the street to visit with their friends in New Bedford or Nantucket, so they now rode from ship to ship in the whaleboats whenever the weather permitted—and when the boats were not whaling. When they could not visit, they stood atop the afterdeckhouses and waved their handkerchiefs at one another—"Mrs. Gibbs and myself had another flourish of pocket handkerchiefs," Mary wrote one windy afternoon. On calmer days the families met for tea or dinner. They looked at one another's daguerreotypes. Everyone took presents to everyone else. There was a constant exchange of raisins and preserves, honey and coconuts, cookies and kittens, pigs and figs, yarn and silk, shells and green tea, salted cod and pickled mullet, books and currant wine.

On one occasion Mary took laudanum for a toothache and went to bed—just as the Addison was hailed by yet another whaleship. "I did not feel very glad to hear it was the Augusta," Mary confessed, "and that Captain and Mrs. Taber were coming on board. However, I got up and fixed up as much as possible at so short a notice, and it did me a great deal of good—cured me completely."

Gamming was also a tonic for the children, who normally missed the company of other youngsters. When Minnie Lawrence and her mother visited his ship, little Charlie Baker of the William Gifford was so happy to see another child his age that he broke into tears as they were leaving. Minnie's favorite was Sammy Winegar. Her mother wrote that "every time he comes on board she asks me if I don't wish he was her brother."

So great was the congregation of whalers in arctic waters that there were times when not everybody could fill his oil casks to capacity. In

Ned Penniman went a-whaling in 1878 when he was eight years old. His father, Captain Edward Penniman, wrote that he "hauls on the ropes like a sailor."

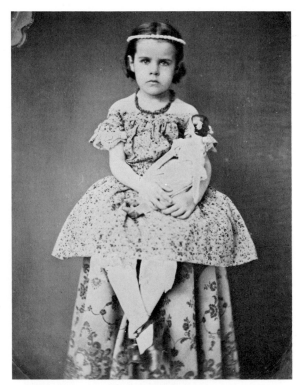

As a six-year-old, Minnie Lawrence distributed Bibles to the Addison's crew in 1857 and, inspired by a Biblical tale, once commanded a gale, "Peace, be still."

Few youngsters could claim such attire as James Smith, who wears a penguinskin coat made by the Chelsea's crew on an 1837 voyage to the Kerguelen Islands.

some years the herds seemed smaller than in others—or perhaps the whalemen missed the main migration—in which case the complaints would be long and loud. Every year the members of the fleet watched with growing anxiety for their quarry to appear.

"Thirty ships in sight within a few miles of Cape Lisburne," Mary wrote. "Is it any wonder that we do not get oil?" During their second cruise north, in 1858, Mary lamented, "Oh, that we might get some oil!" Next day she wrote, "A whale, a whale, a kingdom for a whale!" Eleven days later she was so discouraged that when a whale was chased and lost she broke into tears, a rare occurrence for her. That same day, the *Addison*'s men finally did catch a bowhead, and she announced the event in her journal with "Eureka! Eureka!"

The whales, it developed, were late in their migration that year. When they did appear, the *Addison* stowed away 700 barrels of oil and 10,000 pounds of bone, a good haul compared with that of many other ships, but far less than Captain Lawrence had hoped for.

The next year was even worse. In midsummer Mary was writing, "I cannot think for a moment that Dame Fortune will permit us to leave the whaling grounds at the end of the season without one drop of oil. I exert all my powers to keep up the spirits of the captain and officers." A few days later she recorded the sorry list of other hard-luck whaleships: "The *Good Return* had taken nothing, the *Speedwell* nothing, the *William Gifford* nothing, *Cleone* nothing, *Robert Edwards* nothing, *Arab* nothing, *Christopher Mitchell* nothing, etc. Misery loves company, and it is comforting to know that we are not alone." By August 10 the *Addison* was reduced to chasing walruses for their relatively small amount of oil. The results were disappointing: when a herd was sighted, the boats could not get through the ice, and the men returned with only two of the animals. In her disgust Mary decided that walruses "are the worst-looking creatures that I ever saw, without exception."

Again it turned out that the bowheads had merely been late, and on August 15 the *Addison*'s men brought their first one alongside. To make up for lost time, Captain Lawrence decided to stay later than usual, and his wife worried about reports of ice forming early in the Bering Strait. "It alarmed me a little," she wrote, "fearing that we might get caught in here." Captain Lawrence hung on into late September. The whales that appeared were easily frightened and escaped under the increasingly heavy ice packs. The *Addison* had only 460 barrels of oil and about 6,000 pounds of baleen when Captain Lawrence finally sailed south.

In December of 1859 the *Addison* headed home at last. On May 12, 1860, sailing up the Atlantic, the captain's wife wrote in her journal, "I shall feel badly, after all, to give up my *Addison* home." When the whaler anchored in the Acushnet River on June 14, Mary irritably noted that "the pilot could give us but very little news. He had not a paper on board and did not even know who were the candidates for the presidency." Finally stepping ashore in New Bedford, she was surprised to see that in four years her clothes and Minnie's had become old-fashioned and that hoop skirts were no longer worn. Undoubtedly, though, Mary Lawrence of the *Addison* was pleased to see that corsets, stiffened by baleen from the arctic whaling grounds, were still very much in vogue.

Folk art from the scrimshander's blade

"We are regularly cruising with not enough to do to keep a man off a growl," wrote William Davis in his journal on January 23, 1874. And then he added, "As this habit cankers the soul, I prefer to scrimshone." Whether afloat or ashore, Yankee whalemen occupied their otherwise demoralizing idle hours carving whale teeth and bone into trinkets and other objects falling under the general name of scrimshaw—or as it was also known, scrimshone, scrimshorn, scrimshonter or scrimpshong.

Proclaimed by some as the only truly indigenous American art form, scrimshaw may have originated with the Indians and Eskimos. But the New England whalemen probably discovered it for themselves by putting together two items they had plenty of—time, and whale teeth and bone. Their creations, as can be seen on this and the following pages, ranged from the simplest clothespin to dazzlingly complex gadgets and beautifully decorated *objets d'art*.

The scrimshander, as a practitioner of the art was called, could work in a wide variety of materials. The teeth and jawbones of the sperm whale were most favored, but walrus and narwhal tusk, porpoise jaws and baleen from the mouth of the right, bowhead, humpback and gray whales were also highly regarded.

Usually a jackknife was all that was needed for a carving tool; however, Herman Melville noted that some of the more devoted artists aboard his ship possessed "little boxes of dentistical-looking implements especially intended for the scrimshandering business." The designs that ornamented many of the pieces were usually inscribed with a sail needle and then darkened by rubbing in a mixture of oil and lampblack.

Rings, rolling pins, bird cages, butter knives, doorstops, cribbage boards and minutely exact ship models were common subjects. Of them all, the most popular was the jagging wheel, an instrument used to crimp the edges of a piecrust before it was put in the oven. These, along with busks—thin slats made of baleen, used to stiffen women's corsets—were produced by the thousands by homesick and lovelorn sailors for their wives and sweethearts in New England. Sometimes these gifts were engraved with samples of versification framed by the whaleman to win favor in his lady's heart, such as this ditty from a corset busk given to a New Bedford lass:

Accept, dear Girl this busk from me;
Carved by my humble hand.
I took it from a Sparm Whale's Jaw,
One thousand miles from land!
In many a gale, has been the Whale,
In which this bone did rest,
His time is past, his bone at last
Must now support thy brest.

The teeth of a sperm whale, set in a cartilaginous strip— the animal's gum—that becomes bone-hard after it has dried, are removed from the jaw by a team of men hauling on a block and tackle in this sketch from a whaleman's journal of 1855.

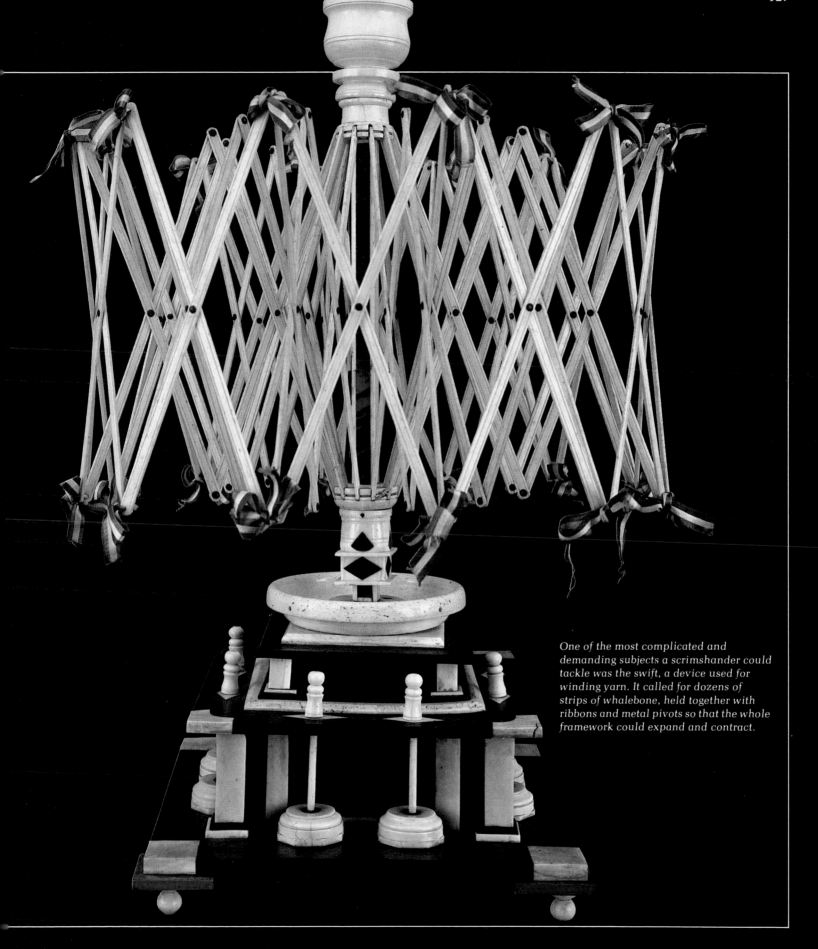

One of the most complicated and demanding subjects a scrimshander could tackle was the swift, a device used for winding yarn. It called for dozens of strips of whalebone, held together with ribbons and metal pivots so that the whole framework could expand and contract.

Jagging wheels for crimping piecrusts

Rolling pin

Spool holder that dispensed thread through
ivory hearts, diamonds and rings set in mahogany drawers

Ditty box for sailor's valuables

Corset busks made of baleen and elaborately
decorated with whaling scenes and fanciful designs

Yarn basket

Bodkins used for embroidery
and other work with thread or yarn

Clothespins

Eggcup

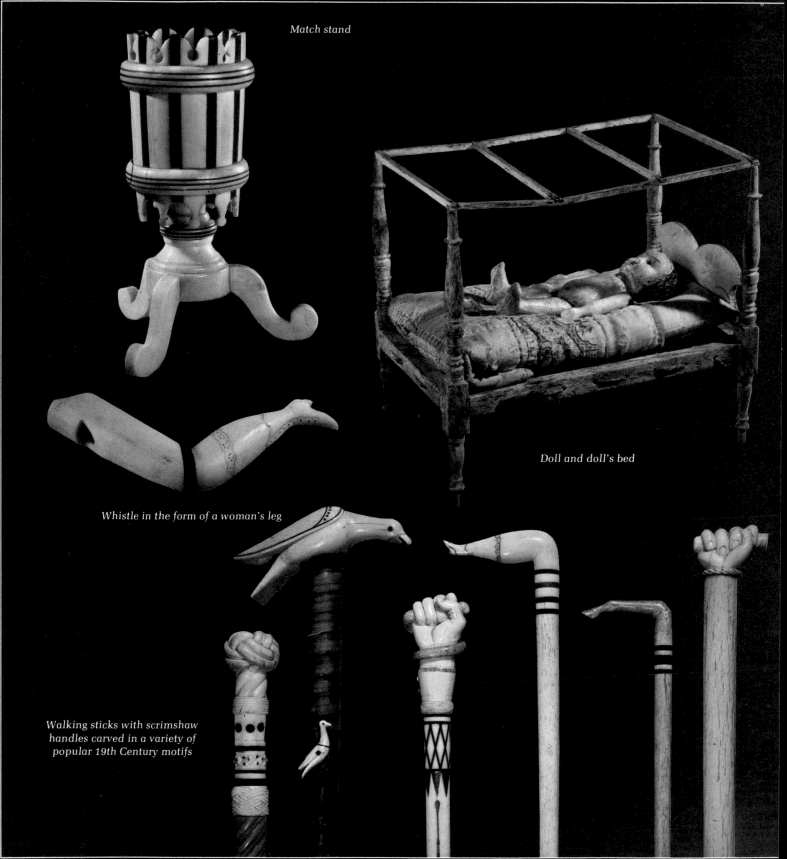

Match stand

Doll and doll's bed

Whistle in the form of a woman's leg

Walking sticks with scrimshaw
handles carved in a variety of
popular 19th Century motifs

Wall pocket for
bedside reading material

Lantern

Miniature whalebone sled only 10½ inches long

Candlestick

Chapter 5

The devilfish of Baja California

Yankee whalers crowd in for the slaughter of gray whales in 1858 at a lagoon in Baja California, once a sanctuary for the animals.

old! came the cry, and the hysteria that swept across the United States in the spring of 1849 coursed like an epidemic through the Yankee whaling fleet as well. Gold in powder, gold in flakes, in grains, in great heavy nuggets, whole riverbeds full of glittering gold had been found around Sutter's Mill in California. To the whalemen setting out from New England or rolling along the Pacific swells, a whole new adventure beckoned, and for a time the greatest of history's hunters virtually abandoned their singular pursuit of the whales.

Captain John N. Smith of the *Garland*, outbound from New Bedford in June 1849 but distracted by all that gold in California, thought that he was a "fool to go a-whaling; that he should not have gone but for the solicitations of his wife, and that he ought to have gone to California." Which is precisely what he did. After rounding the Horn, Smith took the *Garland* not to the Pacific whaling grounds but to San Francisco, planning to run the ship aground and "make a public house of her"—after winning his pile in the gold fields. The *Garland*'s owners rescued their vessel, but without captain and crew, who became prospectors.

The crew of the New Bedford whaler *Inez* learned of the gold rush from another ship while cruising off Australia. They persuaded Captain William L. Jackson to put into Sydney, sell their oil and sail for San Francisco. The crew of the *Henry Clay*, whaling near the Galápagos, put a similar proposition to their captain, Samuel P. Skinner; when he refused, they burned the ship and took off in her whaleboats for the nearest port to find passage to the gold fields. The captain's fate is unknown.

Yet another master, Charles Eldridge of the *Popmunnett*, in Valparaiso for provisions, went on deck one morning in 1850 to find a silent ship. He, his first mate and one foremasthand were the only souls on board. The rest had jumped ship for California; no doubt the foremasthand would have been among them had he not been ill.

The effect of such massive desertions was devastating to the whaling industry. Almost 435,000 barrels of whale oil were landed at New Bedford and other New England whaling ports in 1847; the total plunged to 290,000 barrels in 1850. Eventually, of course, after the gold rush had run its course, whaling, like other industries, quickly recovered. But while the gold rush lasted, a number of shipowners went bankrupt, and times were hard among the workers in the whaling chandleries, oil processing plants and candle factories.

Yet for all the troubles it caused, the California gold rush did have one curious side effect on whaling that was not entirely deleterious. The mania for gold was directly responsible for opening up a new and dramatic form of Pacific whaling that for its brief duration provided one of the most fascinating chapters in the history of the industry in the 19th Century. To a large extent, this development could be credited to the fateful actions of one New Englander, who was among the hundreds of ambitious Easterners who hurried west to make their fortunes.

That Charles Melville Scammon ever became a whaling captain was entirely a matter of accident. He was born on May 28, 1825, the son of a Methodist minister in Pittston, Maine, on the Kennebec River. One brother, Eliakim, became an Army officer, and among his subordinates were such promising youngsters as James Garfield and William McKinley. Another brother, Jonathan, went west for a successful career as a lawyer, financier and civic leader in Chicago. For a while Charles stayed home, read voraciously and wrote poetry for his invalid sister. But at 17, an advanced age for a New England boy, he went to sea. At 23 he was a skipper of a merchant ship. In 1850, at 25, he sailed for California, not so much for gold as for advancement; he assumed, correctly, that there was more of a demand for merchant captains out there than at home.

For two years Scammon had no trouble finding merchant-ship commands, to North Pacific ports, to Chile and to China. By 1852, however, the gold fever was subsiding. San Francisco's needs were not so great, and fewer merchant ships were sailing. So, as Scammon explained it, "the force of circumstances compelled me to take command of a brig, bound on a sealing, sea-elephant, and whaling voyage, or abandon sea-life, at least temporarily." His first vessel was the *Mary Helen*, and it was the beginning of an 11-year stint as a sealing and whaling captain that brought Charles Scammon immortal fame.

His whaling voyages took him far out into the Pacific, up to the Okhotsk Sea and down along the coast of Baja California. And it was while he was off Baja California that he learned the wily ways of the gray whale. Scammon, in fact, became a student of the California gray, and eventually wrote a learned treatise about this intriguing animal.

The California gray whale annually undertook one of the most impressive migrations of any creature, ranging nearly 12,000 miles round trip from the Baja California coast to the Arctic Ocean. During the summer months the grays spent nearly all their time feeding on amphipod crustaceans scooped up from the shallows of the Bering and Chukchi Seas. They dredged the bottom, shoveling up tons of silt, sand and water.

Shipping of every description jams San Francisco Bay during the California gold rush in this 1851 lithograph. The lure of the gold fields was so strong that many a whaleship's crew deserted, and numbers of captains found it more profitable to carry fortune seekers west than to pursue the whales in the Pacific.

Forcing massive tongues against baleen jaws, the grays pushed out the water, and in great gulps swallowed the amphipods into their three-chambered stomachs. By autumn the grays had developed a heavy blanket of blubber and were ready for the southern leg of their migration.

Slipping between the Aleutian Islands and following the northwest and California coastline, the grays headed for Baja California. They swam southward day and night at a steady pace of four knots, rarely pausing for food. Leading were the pregnant females, followed by the females that had either given birth to their young the previous year or were in season for the first time. They were followed by the bulls.

Along the Baja California coast the great herds of grays turned into the lagoons that spread inland along the flat, barren desert. Unlike any other whale, they preferred these shallow, protected bays for mating and giving birth. Most of the pregnant cows moved inland as far as they could go, artfully navigating shallow waters that would panic and strand other species of whales. The grays that were ready for mating remained closer to the lagoon entrances.

The male was usually the aggressor in the gray whale's courtship, approaching and rubbing against the female, who would respond if she was in heat. Because so many females were giving birth, the mating ratio usually was two males to one female, but there was surprisingly little battling between the males. In fact, when a male and female gray had paired off, another male generally stood by as if for protection. And when the male and female grays lay on their sides, the female holding the male in a flipper embrace while they mated, the second male was often seen lying across them, helping to hold them in position.

At the upper reaches of the lagoons, meanwhile, the cows were giving birth much as other whales do, quickly nudging their babies to the surface so they could breathe. The calves were huge in relation to their mothers; a 35-foot gray would have a calf 15 feet long. If the newborn gray was defective in any way, it was immediately abandoned. If not it would be suckled with milk 10 times as rich as that of dairy cows. By the end of winter the young whale already deserved its name: it was patterned with gray like its elders. And it was two-thirds adult size, ready for the long migration back to the arctic feeding grounds.

This life cycle of the California gray whale was little disturbed by whalemen until the mid-19th Century. The graybacks gave much less oil than the larger sperm and right whales, and their baleen was too coarse for use in corsets and buggy whips. Along their migratory route the grays were preyed upon from shore by Eskimos and Indians, with virtually no effect on their population, which was estimated at 25,000. But in the late 1840s a number of captains turned from the arduous long-distance pursuit of the sperm whale to the close-at-home grays. And to their everlasting surprise, it was at first an almost even contest.

The initial attempts against the grays took place in such wide, easily accessible lagoons as Magdalena Bay, two thirds of the way down the Baja California coast. And the attackers were stunned to discover that a gray whale, cornered in its lagoon, could be even more ferocious and destructive than a maddened sperm whale. The gray whales also demon-

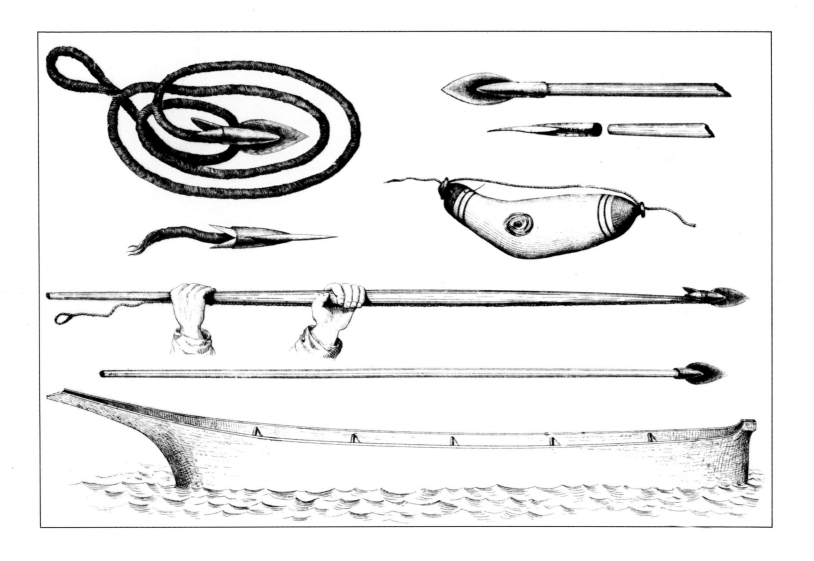

The whaling tools of the northwest Indians are depicted in drawings by the indefatigable Charles Scammon, who in 1874 wrote a learned treatise on west coast whaling. Immediately above the cedar whaling canoe is a killing lance, and above that a harpoon. At upper left are two views of an abalone-shell harpoon head with line attached; at upper right are two views of a lance head, and a sealskin bladder used to mark the whale's position.

strated wariness and a wily intelligence as well as the ability to make the most of the treacherous channels and shallows of the lagoons. This form of whaling became known as "mudholing" because of the terrain, and the grays earned the name of devilfish for their fierce retaliation.

One demoralized mate returned to his ship along the beach and reported to the captain that there was not "enough left of the boat to kindle the cook's fire." Angrily protesting against this new type of whaling, he complained: "I shipped to go a-whaling. I'd no idea of bein' required to go into a duck pond to whale after spotted hyenas. Why, Cap'n, these here critters ain't whales." And what are they? the captain asked sarcastically. The mate replied, "I have a strong notion that they are a cross between a sea serpent and an alligator."

A popular tale of the time described an enraged gray whale pursuing a whaleboat down a channel and onto a sandbar. The captain who spun this yarn claimed that his harpooneer called out as the men scrambled onto the beach, "Cap'n, the old whale is after us still." The captain dryly recounted, "I told all hands to climb trees."

Charles Scammon had heard such stories in San Francisco, but they did not deter him from deciding to add a bit of mudholing to his itinerary during the winter of 1856-1857. In November he took the brig *Boston* into Magdalena Bay, and quickly found that the grays deserved their frightening reputation. Adding up the damage, he later wrote that "two boats were entirely destroyed, while the others were staved 15 times; and out of 18 men who officered and manned them, six were badly jarred, one had both legs broken, another three ribs fractured, and still another was so much injured internally that he was unable to perform duty during the rest of the voyage. All these serious casualties happened before a single whale was captured."

But Scammon and his men stubbornly persisted. "The contest with the 'Devil-fish' was again renewed with successful results. Several whales were taken without accident, and no serious casualty occurred during the rest of the season." Scammon left Magdalena Bay in May of 1857 and sailed for San Francisco with some 500 barrels of oil. For all the problems and hazards, it had been a highly successful expedition. What is more, Scammon had heard from a local Mexican of another redoubt of the grays—one that no American whaleman had ever visited. The man told him of a lagoon in the Vizcaíno Desert in northern Baja California that was jammed with gray whales but was hidden and virtually land-locked. The Mexicans called it Laguna Ojo de Liebre, which translates roughly as "Jack Rabbit's Spring Lagoon." Scammon made a mental note of this interesting information and returned to San Francisco.

The next summer and fall, in the brig *Boston*, he was whaling in the Pacific with what he called "ill success." In fact, by October the *Boston* had laid down only 40 barrels of seal oil and not a drop of whale oil. Scammon remembered the tip he had picked up the previous winter, and decided to try his luck one more time with the devilfish.

However, his crew had signed on for a cruise of only eight months. So Scammon put a proposition to his whalemen: if they would extend their agreement he would try to find a gray whaling ground never before visited by a whaleship. Only three men declined. They were put ashore at Santa Barbara. A messenger was sent to San Francisco with a request for a shallow-draft schooner, the *Marin*, to be brought south. Brig and schooner met in Catalina harbor, and in November of 1857 Scammon took both vessels to Baja California and into history.

Nearly halfway down the peninsula of Baja California, the coastline made an abrupt hook to seaward. Its outer arm no doubt once curved northward to embrace a large bay, but the ocean now broke through between the point of land and two large islands that still defined what was called Vizcaíno Bay. With a fetch of thousands of miles, the Pacific swells crashed against this arm of land and the islands. Swirling currents surged around the islands and into Vizcaíno Bay. At the southern head of the bay the sea, driven by strong prevailing northwesterlies, broke into long, rolling combers that pounded onto the white beaches.

In spots, wave and tidal action had broken through the dunes and had flooded the wide alluvial plain inland. At least one of these break-throughs, according to Scammon's informant, now led into a huge secret lagoon where hundreds, even thousands, of gray whales, threading

Yankee whaleman Charles M. Scammon appears prosperous and convivial in this carte de visite, a photograph that he had printed and distributed to friends and acquaintances in the 1860s after pioneering an entirely new method of shallow-water whaling in the Baja California lagoon that bears his name.

through the channel, took refuge for mating and birthing every winter.

In late December, Scammon anchored near the head of the bay. Looking shoreward, he could see nothing but uninterrupted surf. Crackling, booming and hissing, the sea arched over a barrier sandbar and onto the beach beyond. It was like an impenetrable reef. But somewhere, as with most reefs, there should be an opening. Scammon's Mexican informant had admitted that he had never visited this lagoon, but he had described it in convincing detail. Scammon determined to find out next morning if the man knew what he was talking about.

After a night of rolling in the swells and listening to the thunder on the shore, Scammon dispatched his schooner, led by three whaleboats, to search for the opening. Pitching like corks and struggling to keep from being swept into the line of surf, the boats and the schooner moved away, parallel to the beach. Soon they were out of sight.

That night was an anxious one on the *Boston*, as were the next day and the next night, with no sign of the whaleboats or the schooner. On the third day, with vast relief, Scammon saw one boat returning to the brig. The boatmen had found a channel and had gone through it, sounding as they went and riding the high waves as if on giant surfboards. The channel was deep enough for the schooner to follow the whaleboats through—into a glass-smooth green lagoon stretching inland as far as they could see. The channel, though dangerously narrow, appeared to be more than 30 feet deep, enough for the brig as well. The schooner and the other boats were waiting for the *Boston* to join them inside the lagoon.

Scammon raised anchor and hoisted sail. By midafternoon the *Boston* was moving along the surf line, under a brisk breeze, toward the lagoon's entrance. Right at the channel mouth the waves and currents mixed in a cross chop that sent eddies in every direction. Across the white manes of the breakers, Scammon saw his schooner, her bow pitching into the seas and her masts gyrating as she sailed partway back through the channel to meet him. He ordered the brig's helm put over, and the *Boston* went planing the surf tops into the channel. Then the wind died.

The brig and schooner had almost met, in the shallowest part of the entrance, over a sandbar that thrust the inrushing water upward into breaking combers. The waves and currents threatened to sweep both the *Boston* and her tender onto the shoals edging the channel. Scammon had no choice but to anchor both vessels in the wildest part of the channel and wait for a breeze so they could maneuver out of danger.

No wind came. Darkness and fog descended. The swells rolling in through the channel increased in height, sending waves over the bows. Scammon hoped for a breeze off the land, a common nighttime occurrence in areas like this. At least it would help him escape back into the bay so he could try again next day. No breeze came from any direction. The two vessels pitched and rolled, their yards and booms slashing from side to side and their bows yanking at the anchor lines.

"Not a soul on board slept during that night," Scammon later recalled. "A light puff of wind, at long intervals, came through the mouth of the lagoon, each time giving us hope for the desired landbreeze. But it only increased the dismal sound of the angry surf as it beat upon the sandy shores."

With dawn came a light breeze, and Scammon ordered both vessels under way. Before the anchors could be raised, the breeze died. Rolling and pitching, studying the desolate landscape shimmering under the sun, Scammon and his whalemen waited for wind. They were not encouraged by the sight of a shipwreck on the beach near the channel.

At noon, however, a brisk wind came in from the north. Anchors were quickly heaved, sails were loosed, and the *Boston* and her tender raced through the channel into the smooth anchorage inside the broad lagoon.

But where were the whales? Under the hazy sun the placid water stretched away from the entrance, unbroken by any sign of activity save the wheeling and diving of countless birds. Had the migrating grays not found this lagoon? Certainly they had not been driven out by whalemen. Scammon was sure that the *Boston* was the first whaler to find this hidden refuge. He could only hope that the whales were farther up in the lagoon, or that he was early; it was still December. In the meantime they would have to explore the channels between the sandbanks that patterned the lagoon. But first the *Boston* needed wood for her galley and tryworks. And thereby the expedition almost failed at its start, in a tragicomic episode that Scammon characterized as a day of disasters.

He remembered the shipwreck on the shore, and now planned to turn the other skipper's misfortune to his own advantage by salvaging his needed firewood from the wreck. Sending the schooner and one whaleboat off to start sounding the lagoon, he took the *Boston* and the rest of the boats as close to the beach as he dared. The anchor was sent down and everyone climbed into the boats to row for shore. The plan was to walk across the beach to its outer side, bring wood back from the shipwreck, then ferry the wood out to the *Boston* in her boats.

Scammon later described what happened: "All the boats engaged in transporting it were moored near the shore in the lagoon, and left in charge of a boat-keeper, it being impracticable to haul them up at high tide on account of the broad, flat beach exposed at low water. All the wooding party being out of sight when at the wreck, the boat-keeper concluded to refresh himself by a bath, and conceived the idea of converting one of the boats into a bath-tub, by pulling out the plug in the bottom. The boat soon became water-logged, and the fellow, carelessly enjoying his ablutions, got too far to one side of the craft, which instantly capsized, turning him into the lagoon. The current running swiftly, dragged the anchor, and the man, in his fright, swam to the shore, abandoning his boat, to which three others were fast."

The overturned whaleboat still had enough buoyancy to float and, with the other three boats, it drifted with the current through the channel into the surf and crashed against the outer beach. "The alarm was given to the party on shore," Scammon recalled, "and it was a disheartening sight to behold the four boats drifting through the breakers, for everyone knew that without them our voyage would be fruitless."

Some of the South Sea islanders thereupon resorted to an ingenious solution. Surfboarding had originated in their homeland. Wrenching long planks from the shipwreck, they waded into the surf, lay on their planks and paddled out through the breaking waves. They were joined by the *Boston*'s carpenter, who was an excellent swimmer. The islanders

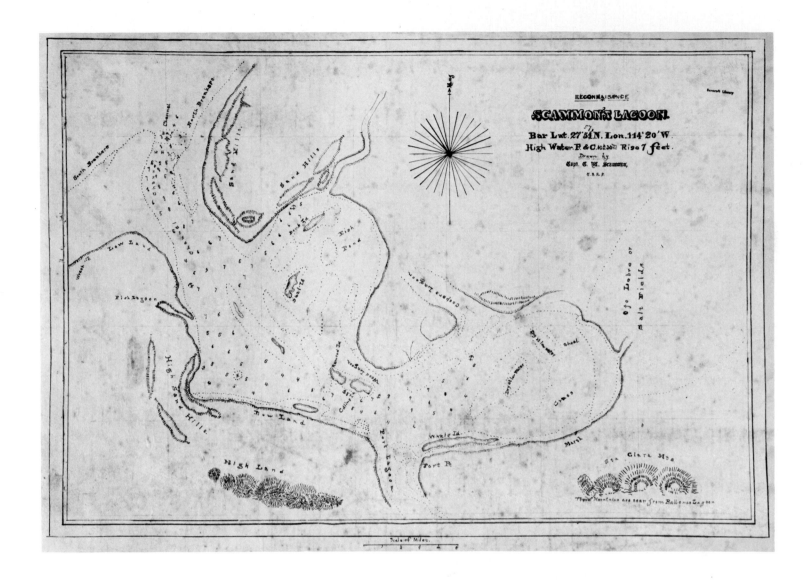

The barrier sandbar and the narrow channel opening into Scammon's Lagoon appear in the upper left corner of this highly detailed hydrographic survey made by the master whaleman, Charles Scammon, himself. He marked the lagoon's depths in fathoms, drew anchors where his ships found good anchorages and used an "X" to indicate rocks; sand hills and highlands on shore were shown in relief.

on their boards and the carpenter churning through the surf seemed to be gaining on the drifting boats when another mishap occurred.

As the boats drifted out to sea, the anchor, which had been bouncing along the bottom, suddenly caught and held. The islanders and the carpenter, moving on a converging course, tried to turn back toward the anchored line of boats. But a current sweeping along the shore was too strong for them. The islanders saved themselves by paddling and riding their surfboards to shore. But the carpenter disappeared.

By this time the tide was turning. "We had a faint hope," Scammon wrote, "that the change of the flood would bring some of the boats, even in a wrecked condition, back or near to the passage of the lagoon." He stationed lookouts along the beach, and shortly "a hawk-eyed young-ster, who had climbed on a sand-hill," reported that a boat was drifting toward the beach. A dozen men wrestled it ashore before the waves could break it up. The surging seas had parted the line holding the boats together; two others came tumbling toward shore and were also saved.

The fourth boat never reappeared. Nor did the body of the carpenter.

Scammon wrote that "it was late in the night before we returned to the brig, tired and dejected over the day's disasters, involving the loss of a favorite shipmate." But there was one consolation. All around were spouts and splashing bodies. The gray whales had reached the lagoon. Scammon and his crew had simply arrived a few days too early.

Scammon never reported what punishment was meted out to the hapless whaleman who had almost cost them their boats; perhaps it was better left unsaid. It took three days to get the three remaining whaleboats back over the wide beach and into the lagoon, load them with wood and bring it out to the *Boston*. At that point a northerly gale swept down the coast and enveloped them. Finally the weather improved and, led by the schooner, whose men had been charting the lagoon, Scammon took the brig some 30 miles inland to the inner reaches of the bay.

Here were even more whales, rolling, spouting, diving, chasing one another, sending up billows of mud as they plowed the bottom, and "spy hopping"—raising their long barnacled bodies partway out of the water, turning in a circle and falling on their sides. Scammon and his men were sure that the whales were studying the intruders—and they were at least partly right; because they spent so much time in shallow, muddy waters, gray whales were the most visually oriented of the great whales and often rose up to survey their surroundings.

The *Boston's* men lost no time in lowering away, and soon had two gray whales alongside, ready for the tryworks. At last it looked as if Scammon's gamble was about to pay off. Next day, however, the grays once again confirmed their reputation for combativeness. Two boats set out. One was nearly on a whale when the great animal, as if craftily waiting for the men to close in, raised its giant flukes and brought them smashing down on the boatload of men. The boat was reduced to kindling, and it is a wonder the six men were not killed. As it was, two suffered broken limbs and another three were injured less seriously. The second boat went to their rescue, and was attacked and stove in by another whale, with more injuries to the whalemen. The schooner, being closest to the scene, sent her ship's boat to pick up the floundering men.

When this boat came alongside the *Boston*, Scammon recounted, "it could only be compared to a floating ambulance crowded with men." Not only were nearly half of Scammon's men disabled for days but the uninjured whalemen were utterly demoralized by the whales' attack.

Gradually their terror subsided and, Scammon wrote, "two boats' crews were selected and the pursuit was renewed. The men, on leaving the vessel, took to the oars apparently with as much spirit as ever; but on nearing a whale to be harpooned, they all jumped overboard, leaving no one in the boat, except the boat-header and the boat-steerer."

Scammon was nearly disconsolate. "Our situation was both singular and trying," he reflected later. "The vessel lay in perfect security in smooth water; and the objects of pursuit, which had been so anxiously sought, were now in countless numbers about us. It was readily to be seen that it was impossible to capture the whales in the usual manner with our present company."

The irony of his situation, he realized, was that, while it would appear that he had trapped his quarry in its refuge, the terrain in fact favored the

Hunting gray whales at the mouth of San Ignacio Lagoon in Baja California, a harpooneer in the whaleboat at center prepares to fire a bomb-lance gun, which contains a deadly projectile that would explode after penetrating the mammal's body. The sketch is by Charles M. Scammon, who found the bomb-lance gun especially useful in waters where there was little room to maneuver and the whales were particularly dangerous.

whales, and they were using it to advantage. In the open sea a sperm whale often tried to flee when struck by the harpoon. But here in the lagoon the cornered gray whale turned to fight. Moreover, Scammon was discovering, few creatures in nature were as aggressive as a female gray whale in defense of her young. As in previous encounters with grays, the cows were the chief attackers. And in the channels of the lagoon the whales could easily outmaneuver the most skillful oarsmen. It began to look as if Charles Scammon would have to admit defeat.

But first he decided to try a new tactic. "Among the officers there were two who had been considered good shots with the bomb-lance gun, one of whom we personally knew to be unequalled as a marksman," he wrote. The bomb gun was a rifle-like weapon that came into use in the early 1850s; it fired a projectile that exploded a powerful charge when it penetrated its target. A well-placed shot could kill a large whale.

Yankee whalemen normally scorned this device, relying on their old favorite, the harpoon. A harpoon fastened itself to the whale more securely, they reasoned, and nothing killed so efficiently as a deftly handled lance. But this method required a great deal of room for maneuver, and there was no such room in the shoal-dotted bay. So Scammon turned to the bomb gun, and devised a new method of whale killing.

"The officers were called together," Scammon reported, "and the matter was plainly set before them. The marksmen were informed that if they could kill a whale without expending more than three bomb-lances, our supply was ample to insure a 'full ship.' " The strategy was to position the boats on the side of the narrowest channel in the lagoon, near where the whales passed but in water so shallow that the creatures could not attack the boat. The bomb gunner would wait until a gray whale came within range and then shoot it.

Scammon had three boats ready: two from the *Boston*, including the one that had been stove in but was subsequently repaired, and the ship's boat from the *Marin*, which was pressed into service as a whaleboat. These craft were immediately dispatched, "two prepared for shooting and the third as a relief boat in case of emergency," recounted Scammon. "They took their positions as ordered, and it was not long before three whales had been 'bombed'—the third one was killed instantly and secured. On returning to the vessel, the officers reported their good luck; and on the following day they were again dispatched, but with instructions to first board the tender, and take a look from her mast-head for the whales that had been bombed the day previous, as we confidently expected that either one, or both, would be found dead not far from where they were shot. It was a pleasant surprise to the chief officer, when, on going half-way up the rigging, both whales were seen floating near the head of the lagoon; and no time was lost in securing them."

From that time on, Scammon wrote, "whaling was prosecuted without serious interruption." Such a method would not have been practical at sea, where a wounded whale could dash out of sight before dying. But in the confines of the lagoon it worked perfectly, said Scammon, "without the staving of boats or injury to the men." In fact, he added, "the try-works were incessantly kept going—with the exception of a day, now and then, when it became necessary to 'cool down' in order to stow away the oil and clear the decks—until the last cask was filled." The bread was removed from its casks and stowed in a bread locker so these casks could be filled with oil. Covers were made for the tubs in which blubber was minced before it was tried out, and these too were filled.

Even when filled to brimming, the *Boston* and the *Marin* could hold only about 700 barrels of oil, the product of 20 whales, before Scammon was forced to call it quits and return to San Francisco. It was a modest catch considering the abundance of whales, and a far cry from the thousands of barrels of oil returned from the Pacific by larger vessels. But it was the best Scammon could do; had he gone south with larger, more capacious vessels, he could not have entered the lagoon at all and would have returned empty-handed.

At that, he was lucky to clear the lagoon with his cargo. In a light morning air one day in early spring, the *Boston* and her tender hoisted anchor and moved toward the entrance channel. As they reached it, the wind blew offshore straight down the channel, which was too narrow for the *Boston* to tack upwind. It was the season for northerlies, and for two maddening weeks gale after gale raged into the lagoon, while the two vessels swung at their anchors and Scammon and his men fretted with frustration. Curiously, for all the wind, there was little rain, and in time the supply of fresh water ran dangerously low. Everyone had to be rationed one pint a day. At last, the winds swung around out of the east, so Scammon could attempt the channel. But now he found that more than a month at anchor had so fouled his ships' bottoms that even with a strong wind "they made their way at a snail's pace through the breakers."

Waylaying the grays from stations on shore

So closely did the gray whales hug the California shoreline during their seasonal migrations that a number of enterprising 19th Century whalemen simply hunted their quarry from whaling stations on shore—resurrecting a practice that went back to the earliest days of whaling.

A onetime New England captain by the name of J. P. Davenport set up the first California shore station in Monterey in 1854. With tryworks and storehouses conveniently located close to the water's edge, Davenport and his 12 crewmen sallied forth to meet the whales offshore, and before long were producing some 1,000 barrels of oil per year. By the late 1860s, 16 whaling stations dotted the Pacific coast from Half Moon Bay near San Francisco to Mexico's Baja Peninsula.

The companies, as they were called, generally included a captain, a mate, two harpooneers and about a dozen seamen to man two whaleboats.

Using a harpoon gun to fasten to the whale and an exploding bomb lance for the kill, the crews usually made short work of their prey. They then towed the carcass back to the beach, "where," as described by whaleman Charles Scammon, "try-pots were set in rude furnaces, formed of rocks and clay, and capacious vats were made of planks, to receive the blubber."

Without the expense of a whaleship, these outposts at first proved highly profitable. Some shore crews landed 25 whales a year—about the same number as a whaleship's catch. In all, the whaling stations grossed more than one million dollars in the first 22 years of operation, and in that time, according to Scammon, "not less than 2,160 grays, and 800 humpbacks and other whales were robbed of their fatty coverings."

But shore whaling had problems as well. Sometimes the whaleboats had to venture as far as 10 miles out to sea before they found the migration stream, and in many cases were unable to bring home the whales they killed. Almost 20 per cent of all whales killed by shore crews were lost because of bad weather, or because the carcass sank during the long tow to land.

The heyday of shore whaling lasted not much more than three decades. The combined efforts of land-based companies and whaleships resulted in an overkill that reduced the population of California gray whales to only a small fraction of its former numbers. Although a few diehards struggled on into the 20th Century, by the late 1880s most shore whaling stations had been abandoned as unprofitable.

Smoke from the roaring tryworks smears the sky as California shore whalers strip the blubber from their prize and haul the pieces ashore in this 1877 woodcut. In the inset at upper left a lookout scans the sea from an elevated perch while bracing himself against the flagpole, which will display his signals to the boats offshore.

Nevertheless, out they went. The *Boston*, so laden with oil that her scuppers were awash, scraped bottom upon reaching the bar at the entrance. Lurching and bumping, heeling under the wind, she limped over the bar. Looking astern, Scammon could see sand swirling in his wake.

At last in open water, he shaped a northwesterly course across Vizcaíno Bay for Cedros Island, where there would be fresh water. Another gale screamed down on them, and "a heavy, breaking sea continually washed over the vessels, from stem to taffrail." Through a wild night they fought their way to the island, anchoring as even stronger winds came on. For two more days the vessels plunged at their anchors; no one could go ashore, and the fresh-water supply virtually disappeared.

When the gales had subsided, the parched whalemen hurried ashore and luxuriated in the water. It took several days to repair damage and ready the ships for the nearly 900-mile voyage to San Francisco. This time the gales held off, and in March 1858 Scammon brought his rich haul in past the Golden Gate. "Thus ended," he wrote in a masterly understatement, "a voyage which in no small degree was a novel one."

Charles Scammon and Scammon's Lagoon became a whaling legend after that novel voyage. One popular version claimed that Scammon kept his discovery a secret for many years, filling his ship every winter until some of his competitors found his lagoon by following him to it.

In fact, while Scammon could hardly be expected to gossip about his great discovery in every whaleman's bar in San Francisco, he was not overly secretive. To most questioners envious at the sight of his heavily laden ships, he said that he had been to Cedros Island, which was true up to a point. He did give the exact location to his brother-in-law, Captain Jared Poole of the bark *Sarah Warren*. Meanwhile Scammon's mate passed the news to his brother, a whaling captain, who told everyone in the Sandwich Islands. And so next winter, when Scammon took the bark *Ocean Bird* and two schooners back to his lagoon, he found that he was far from alone. "A large fleet of ships," he wrote, "hovered for weeks off the entrance, or along the adjacent coast." Eighteen vessels had come from the Sandwich Islands alone. But only five of the captains managed—or dared—to follow Scammon and Poole over the breakers and through the narrow channel. No doubt many were deterred by the recent experience of the whaler *Black Warrior*, which had been driven onto the shoals and wrecked. But the whaleships that did get through the passage were enough to doom the gray whales of Scammon's Lagoon.

Every day at least 25 whaleboats crisscrossed the lagoon. Most of the whale hunters followed Scammon's example in using the bomb-lance guns. But most were too impatient to employ his tactics of ambush and instead pursued the whales all over the lagoon. Scammon did not ponder the fate of the whale. His vivid description of the carnage at the head of the lagoon is that of a Victorian whale hunter pure and simple: "Here the objects of pursuit were found in large numbers, and here the scene of the slaughter was exceedingly picturesque and unusually exciting, especially on a calm morning, when the mirage would transform not only the boats and their crews into fantastic imagery, but the whales, as they sent forth their towering spouts of aqueous vapor, frequently tinted with blood, would appear greatly distorted. At one time, the upper sections of the boats, with their crews, would be seen gliding over the molten-looking surface of the water, with a portion of the colossal form of the whale appearing for an instant, like a spectre, in the advance; or both boats and whales would assume ever-changing forms, while the report of the bomb-guns would sound like the sudden discharge of musketry; but one can not fully realize, unless he be an eye-witness, the intense and boisterous excitement of the reckless pursuit, by a large fleet of boats from different ships, engaged in a morning's whaling foray.

"Numbers of them will be fast to whales at the same time, and the stricken animals, in their efforts to escape, can be seen darting in every direction through the water, or breaching headlong clear of its surface, coming down with a splash that sends columns of foam in every direction, and with a rattling report that can be heard beyond the surrounding shores. The men in the boats shout and yell, or converse in vehement strains, using a variety of lingo, from the Portuguese of the Western Islands to the Kanaka of Oceanica. In fact, the whole spectacle is beyond description, for it is one continually changing aquatic battle-scene."

The slaughter was in fact so great that most of the gray whales in Scammon's Lagoon were killed or scared away in those first few years. An estimated 550 of the animals were taken. By the winter of 1863 the lagoon was so depleted that it attracted only three vessels; they caught seven whales. And when a Captain Kelley of the *Nile* worked his way into the lagoon in 1865, he found no whales whatever; he became so despondent that "in a fit of mental aberration" he threw himself overboard and drowned. He was buried on the beach.

Scammon's Lagoon was not the largest refuge for gray whales. Magdalena Bay continued to attract whaleships after Scammon's Lagoon was deserted again. But the events in Scammon's Lagoon provided in microcosm the story of gray whaling elsewhere in Baja California. The slaughter had a disproportionate effect because it occurred during birthing and mating season; many cows were killed while still carrying fetuses, and many cows and bulls as they were about to propagate. Within a decade of the time Scammon threaded the channel into his lagoon, gray whaling had become uneconomical. Charles Scammon was 32 when he made his discovery. He retired from whaling at the age of 37. In 1863 he took command of the cutter *Shubrick* in the United States Revenue Marine. He spent another 14 years at sea, including an exploration of Siberian waters and the Bering Sea. But he never lost his interest in whales. He wrote a number of articles on his former quarry and in 1874 published his major opus. *The Marine Mammals of the North-western Coast of North America* remains a classic analysis of the gray and other whales.

There were many ironies in Charles Scammon's adventure. He was a whaling captain by chance, but one of the most successful whalemen of his time. He found a lagoon full of whales, only to be frustrated until he devised his unique method for catching them. But the ultimate irony of his career was that in later years, when naturalists and environmentalists worked successfully to save the California gray whale from extinction, they were greatly aided by the research of the man who had found "the scene of the slaughter" so "exceedingly picturesque."

Chapter 6

The ultimate ordeal by fire and ice

As a rescue whaler stands by, the men of the Acors Barns salvage a few last items from their ship, hopelessly locked in arctic ice in 1876.

t is a safe assumption that in 1859 very few, if any, whalemen had heard the name Edwin L. Drake. And there was no reason why they should have, for Drake was an obscure entrepreneur. But over the next decades every whaleman would come to know, and curse, Drake's name. For on August 27 of 1859, his Seneca Oil Company succeeded in extracting petroleum from the earth by drilling 69 feet down into the soil near Titusville, Pennsylvania. In so doing, he put an end to the Yankee whaling industry just as surely as if he had drilled a hole in the hull of every whaleship.

It took a little time, however. Though petroleum promised cheaper illumination and lubrication than whale oil, its early development proceeded in fits and starts—and a number of extremely loud bangs, when the stuff exploded while being refined. Oil for lighting was prohibitively expensive until the time of the Civil War, when a method of extracting kerosene at reasonable cost was perfected. Nor was petroleum, in those early years, anywhere near the equal of sperm oil in lightness and lubricity. Clocks and watches ran better with sperm oil. The New Bedford *Republican Standard* boasted: "Sperm oil is indispensable for the running of any kind of machinery and all who have anything to do with fine machinery eventually come back to this first principle."

Moreover, the market for baleen, that wonderfully flexible material found in the mouths of all hunted whales except the sperm, continued to grow rapidly. "Bone," as the whalemen inaccurately called it, was increasingly used for corset stays and buggy whips, umbrellas and fishing rods, billiard table cushions and shoehorns.

And so whaling continued to prosper into the 1860s. Sperm oil brought $41.42 a barrel in 1861, and Yankee whalemen shipped 68,932 barrels of it that year. Baleen shipments totaled 1,038,450 pounds in 1861 and at 66 cents per pound earned a handsome profit. New Bedford was now home to more than 500 whalers, which on their long-distance courses into the Pacific spent the winter months hunting sperm whales in the middle latitudes and the summer months on the arctic grounds pursuing the oil- and baleen-rich bowheads.

But now, in the midst of this twilight prosperity before the age of petroleum, the Yankee whalemen were about to undergo two ordeals that would cripple their industry and hasten its demise. The first ordeal was by fire, the second by ice.

Even before the first Yankee whaler was put to the torch by a Confederate cruiser, the American Civil War cut deeply into the whaling industry as irreplaceable seamen and officers joined the naval forces. Massachusetts provided almost 21 per cent of the officers in the Union Navy, and many of them had been whalemen. But it was the Confederate strategy of hunting down and burning Yankee whalers on the high seas that dealt whaling the most devastating blow. This attack on the whaling fleet was all the more demoralizing because it came as a complete surprise.

So far as anyone in New England knew in 1861, the Confederate Navy consisted of one warship, the *Sumter*, a steam- and sail-powered bark with a single 8-inch cannon and four light 32-pounders. They were not yet aware that a secret agent, James Dunwoody Bulloch, had been sent to

Confederate sailors leap into the seas and swim for a lifeboat, as the raider Alabama, long the bane of Yankee whalers, meets her end off Cherbourg, France, after a battle with the U.S.S. Kearsarge on June 19, 1864. As crowds watched from a hillside, the combatants bombarded each other for one and a half hours, until the Alabama slipped stern first to the bottom of the English Channel. At far left, the private yacht Deerhound steams in to rescue the survivors.

Europe by the Secretary of the Confederate Navy, Stephen P. Mallory, to buy several steamships that could be employed as commerce raiders.

At first Bulloch could not find any suitable ships, so he had one built. For £47,500 he contracted with a Liverpool shipyard to construct a 235-foot steel-reinforced bark with a 300-horsepower auxiliary engine. He initially called the vessel the *Enrica* to avoid any suspicious connection with the South, but renamed her the *Alabama* once he had put to sea. By prearrangement she rendezvoused off the Azores with the Southern supply ship *Agrippina*, from which she received her armament of eight cannon, including six 32-pounders. The *Agrippina* also brought the *Alabama* 149 Confederate Navy crewmen under Captain Raphael Semmes, a valiant Marylander who had spent more than 30 years in the United States Navy before defecting to the Confederacy.

In August 1862 the *Alabama* began her career by overtaking the un-suspecting Yankee whaler *Ocmulgee*. The Rebel raider was flying the Stars and Stripes as she came alongside her prey, and only at the last instant did she show her Confederate colors. In short order the *Ocmulgee*'s captain and crew were put into their whaleboats with enough supplies to reach land; then their ship was fired. It was a pattern that would be repeated throughout the North and South Atlantic for nearly two years. In all, 70 Yankee ships of various sizes and descriptions were taken by the *Alabama* before she was finally sunk in the English Channel by the Union warship *Kearsarge* in June 1864. Semmes and most of his crew escaped; the 55-year-old captain returned home to become a rear admiral in charge of a Confederate squadron, but numbers of his men waited in England for a worthy successor to the *Alabama.*

Indeed, the *Alabama's* depredations were soon overshadowed by those of another raider, the *Shenandoah*. In the fall of 1864 that tireless agent, James Bulloch, heard of a swift, rakish transport ship built by Glasgow's prestigious Stevens and Sons. Named the *Sea King*, she was 222 feet long, and ship-rigged with auxiliary steam power and a smoke-stack that could be lowered to make her look like a sailing vessel. Her hull was of wood planks over steel frames. The *Sea King* had just re-turned to London from Bombay when Bulloch arranged to send her to the Madeiras under the guise of a merchant voyage. Here the purchase for £45,000 was completed, and she was renamed the *Shenandoah*. At the same time, according to plan, she rendezvoused with a Confederate ship from Liverpool, the *Laurel*, which carried eight cannon and ample supplies to be transferred to the newest Rebel raider, as well as a crew of more than 100 men, some of whom also transferred to the *Shenandoah*.

Her captain was Lieutenant Commander James I. Waddell, a lanky North Carolinian with thick sideburns, a sun-browned face and a limp from a dueling wound. Like Semmes, he had spent many years in the United States Navy before choosing the Confederacy. He was arrogant, aloof and something of a tyrant. He was also a natural-born sea raider.

Leaving the Madeiras in October 1864, Waddell took the *Shenandoah* around the Cape of Good Hope, put in briefly at Melbourne, Australia, for repairs and then laid a course for the South Pacific whaling grounds around the Caroline Islands. He arrived at Ascension Harbor on April 1, 1865, to find four whalers lying at anchor. Four armed boat crews put out from the *Shenandoah*, and in a trice the whalers were captured. Waddell let the Ascension Islanders loot the ships, after he had stripped them of everything he could use. And while the oil-filled ships burned through the tropical night, the sound of drums and chants carried across the water as the islanders, who had little love for the whalemen because of previous mistreatment by whites, celebrated with a war dance.

Among the items Waddell took from the ships were charts of the Pacific and arctic seas, where the Yankee whaling fleet would be con-centrated during the summer months. Using the whalemen's charts, plainly marked with rendezvous points and whaling grounds, he set his course north for the Okhotsk Sea between Japan and Siberia.

Waddell had scarcely reached the Okhotsk Sea near the end of May 1865 before he came upon the New Bedford whaler *Abigail*. And here he encountered one of the perversities of fate. On April 9, more than a month before, Lee had surrendered to Grant at Appomattox. The Civil War was ended. But Waddell could not know it, and apparently neither did the *Abigail's* captain, Ebenezer Nye. At least there is no evidence to indicate that either Nye or any of his crew made more than the usual protests when Waddell ordered the vessel burned.

While the Confederate raider Shenandoah stands by (center left), hundreds of captured Yankee seamen sent on board the three-masted whaler Milo (right) watch helplessly as smoke pours from the Sophia Thornton and the Susan Abigail, and from three other hapless whaleships on the horizon. The painting is a composite; the ships were actually burned over a two-day period in June 1865, at the apex of a cruise during which the Shenandoah devastated Yankee whaling in the Arctic.

But the *Abigail* exacted her own small revenge—if it could be termed that. In her hold were 25 barrels of whiskey, which the *Shenandoah's* crew naturally treated as contraband of war. The resulting drunken chaos verged on mutiny, and Waddell regained control of the situation only after clapping a large number of sailors in irons.

However, the whaler also provided the Confederate commander with some allies, and one of them was to prove invaluable. Waddell persuaded 17 of the *Abigail's* 36 men to switch allegiance and join the *Shenandoah's* crew. The highest-ranking defector was Thomas Manning, second mate of the *Abigail.* Though he came from Maryland, which remained in the Union, he regarded himself as a Southerner and volunteered to pilot the *Shenandoah* to the whalemen's favorite grounds. Waddell immediately made Manning a corporal of marines and, following his directions, took the *Shenandoah* north and east, from the Okhotsk to the Bering Sea, where the whaling fleet was concentrated.

The Southerners were in the meantime suffering from the arctic weather. Ice formed more than two inches thick on the vessel's rigging, and the sailors climbed aloft to break it up with wooden billets, showering the deck with shards until the ice was ankle-deep. Waddell bundled his head in a shawl, turned up his collar and tied his flowing mustache ends behind his neck. He spent as much time as he could in his cabin, which was heated by a wood-burning stove looted from the *Abigail.*

The *Shenandoah* entered the Bering Sea on June 16, and six days later was off Cape Navarin, where a lookout spotted the telltale evidence of floating blubber. "Steam was ordered," wrote Waddell, and within an hour a man at the masthead was calling "Sail O!" In less than a week, between June 22 and June 27, the *Shenandoah* captured 13 whalers.

The masters and crews of these vessels seem to have been as much in the dark about the surrender at Appomattox as Captain Nye and the men of the *Abigail.* Waddell later recalled that on one of the whalers, the *Susan Abigail,* his men found newspapers reporting that the Confederate government had fled from Richmond, Virginia, to Danville under pressure from Union forces. And when her captain was questioned about the War, recounted Waddell, he responded that "opinion was divided as to the ultimate result of the war. For the present the North has the advantage, but how it will all end no one can know, and as to the newspapers they are not reliable."

Waddell was now burdened by about 250 prisoners, far too many to accommodate on the *Shenandoah.* For a while he tried to tow the shivering crews behind the steamer in a long line of whaleboats. But in the bitter weather that obviously was no solution, and Waddell soon realized that the only course was to put the prisoners on board two of the captured whalers and send them off to freedom and San Francisco. However, he was not happy about it. When one of the captains complained that he lacked enough food for the men packed on board his ship, Waddell replied, "If the rations run out, you can cook your Kanakas."

On June 27 Waddell wrote that "things had gotten lively again." Suddenly there were another 11 ships in sight on the horizon. While the wind held, Waddell lowered the *Shenandoah*'s smokestack and followed his victims, masquerading as another whaler in order not to scatter the prey. It was easy to mistake his steamer for one of the fleet; she was rigged as a three-master and carried her boats on outside davits like a whaler. The next day was calm. Waddell ordered the smokestack raised and the steam boilers fired up, and moved in for the kill.

Nearing her prey, she came on a fraternal scene of whaling cooperation. The *Brunswick* had struck the ice and holed her hull, and the other captains had gathered to offer assistance. In fact, a boat from the *Brunswick* approached the *Shenandoah,* and the unsuspecting mate, evidently assuming her to be a new steam whaler, asked for help. "We are very busy now," Waddell answered, "but in a little while we will attend to you." The mate thanked him and returned to the listing *Brunswick.*

Waddell maneuvered his ship into position to command the fleet with her guns. He then had the Confederate flag hoisted and sent his armed boats out across the water to seize the ships. Instantly 10 of the 11 American flags came fluttering down from their mizzen gaffs.

The one brave flag flying was on the bark *Favorite,* out of Fairhaven, Massachusetts, Captain Thomas Young commanding. A grizzled whaleman in his sixties, Young was a part owner of the *Favorite,* and he was in no mood to see his investment go up in flames—not without an argument. When the Confederates ordered him to haul down his flag, he stuck a bomb gun in their faces and roared, "Haul it down yourself, God

Appearing as a bearded Rip Van Winkle in this 1865 Harper's Weekly cartoon, Commander James Waddell of the Rebel raider Shenandoah puts on a show of astonishment at hearing from a Liverpool harbor pilot that the American Civil War had ended more than eight months before. Exclaims Waddell, fresh in from burning Yankee ships: "Law! Mr. Pilot, you don't say so! The war in America is over? Dear! dear! who'd ever a' thought it!"

damn you!'' Young, in fact, was prepared to die in defense of his vessel. When Confederate boarders approached him on deck, he tried to fire into them. But his weapon misfired, and he was soon overwhelmed. At that, he refused to leave the ship under his own power and had to be hoisted, bellowing and cursing, by block and tackle over the side.

Commander Waddell recorded in his log that Young's defiance had resulted from "too free a use of intoxicating liquor." Nevertheless, he called the stubborn old Yankee "the bravest and most resolute man we captured during the cruise." Young did not return the compliment. He later claimed that his captors had put him in irons and stolen his money, his watch and his cuff links.

Only one other member of the fleet offered any resistance. A cantankerous sow on the whaler *Nassau* had been teased so often by the foremasthands that she had concluded that the best defense was a headlong offense. When the Confederate prize crew climbed to the *Nassau*'s deck, the sow charged, scattering half a dozen Southerners. It took the combined efforts of the prize crew to wrestle her into grunting submission.

The *Favorite*, the *Nassau* and seven other whalers erupted in flames as they were touched off one by one. The black, icy waters were lighted by streaks of flaming oil in what Waddell, with a warrior's view of beauty, called "a picture of indescribable grandeur." The holocaust was punctuated by explosions that the *Shenandoah*'s captain, knowing that whale oil did not explode, said "betrayed the presence of gunpowder" on board the supposedly peaceful vessels. But the gunpowder was only that carried for bomb guns and a few rifles and pistols—standard equipment on any whaler. Waddell was merely looking for justification.

Waddell again had so many prisoners that he had no choice but to spare two more whalers to transport the crews to San Francisco. One vessel he selected was the *James Maury*, whose captain had died at Guam on March 24 on his way to the Bering Sea. His widow, not wanting her husband buried at sea, was preserving his body in a cask of whiskey. When Waddell was told of her presence, with two children, he sent a gallant message stating that the whaler would be spared: "The men of the South do not make war on women and children."

The men from this fleet were the last Yankee whalemen to suffer the indignity and financial disaster of capture by Lieutenant Commander James Waddell and his Confederate sea raiders. Turning south into the Pacific, the *Shenandoah* on August 2 spoke the British bark *Barracouta*, bound from San Francisco to Australia. When an officer of the *Shenandoah* asked for war news, the British captain replied, "What war?"

Told that the conflict had been over for three months, Waddell was at first stunned and unbelieving. At last he was convinced, but now he had a problem. What of his position? He had captured 25 American whalers, had destroyed their cargoes of oil and bone, and had burned 21 of them— all after the cessation of hostilities. Did that make him a pirate? Was an American Naval squadron at this very moment racing to bring him to justice? Would he and his men all be hanged?

Waddell did not linger in the Pacific to await his fate. He took the *Shenandoah* swiftly back around Cape Horn and up the Atlantic to Liverpool, where British authorities seized the ship on behalf of the

United States but allowed the men to go free. In time the *Shenandoah* was sold to the Sultan of Zanzibar for $108,628, about half her cost to the Confederacy. The Sultan planned to use her as his private yacht, but was dissuaded by a frugal royal treasurer, and when the *Shenandoah* ended her days in an Indian Ocean storm, she was hauling ivory and coal.

As for Waddell, he returned home without penalty, and in 1875 was appointed captain of the liner *San Francisco*, sailing between New York and Melbourne for the Pacific Mail Line. He eventually retired to Annapolis, Maryland, where he held court on a bench at the Naval Academy and regaled midshipmen with tales of his derring-do in the Pacific.

All told, the *Shenandoah*, the *Alabama* and other Confederate raiders captured 46 New England whalers during—and after—the Civil War. Another 39 of the whaleships were sacrificed in a bizarre attempt to blockade the ports of Savannah and Charleston by sinking ships loaded with fieldstone at the harbor entrances *(pages 156-157)*. Taking into account other losses from storms, fires, groundings and old age, the War years cost the whalemen nearly 50 per cent of their fleet. Much of the loss was never made up. In America after the Civil War the lure of the West was making fo'c's'le life less and less attractive, and whaling crews were correspondingly inferior. Petroleum prices and quality were becoming competitive with those for whale oil. Emerging industries, particularly Yankee cotton manufacturing, were draining off capital once reserved for whaling. Among the resulting postwar anomalies was the use of whale oil to lubricate the textile mills that were driving whalers out of business and the use of kerosene lamps to light whalers' fo'c's'les.

In only one year after 1865 did United States sperm oil shipments exceed 50,000 barrels—in 1870, when the figure was 55,183 barrels. That was less than a third of the 166,985 barrels shipped in 1843. Now

When whaling ships sank like stones

A Union Navy task force sinks a fleet of whalers in the harbor entrance at Charleston, South Carolina, in this engraving based on an eyewitness sketch.

the more lucrative product brought back by the whalers was baleen. And so more and more whalers were concentrated in the Arctic Ocean, where their quarry was the bowhead whale. And it was in these precincts that the whalemen would shortly face a torment even more devastating than the depredations of Confederate raiders.

From October to June the Arctic Ocean north of the Bering Strait was virtually one solid mass of ice. But as the ice broke into ragged floes and the streaks of water between them widened into meandering pathways, the whaleships gathered to await the arrival of the bowheads.

No one knew where these whales went during the winter. But by June a few usually appeared at the edge of the ice mass, following vast streams of brit, the shrimplike crustaceans that drifted near the surface of the frigid water. The chunky, nearly ovoid bowheads moved north slowly, spending much of their time feeding, or "scooping," as the whalemen described it. Flukes up and heads down, they opened their cavernous mouths and, forcing the water out with their tongues past 500 to 700 fringed baleen plates, they gulped down millions of button-sized brit.

A surface feeder, the bowhead was less ferocious when attacked than either the gray or the sperm whale. Instead it was so timid it was difficult to approach. It could sweep its wide flukes nearly from eye to eye, and the whaleman had to slip in past this thrashing protection to sink his iron. Once hit, the bowhead could dive like a sperm whale, plunging into the mud to "sulk," as the whalemen called it, at the bottom. One bowhead, struck by a Captain Comstock, stayed below for an hour and 20 minutes while Comstock and his crew shivered in their boats. "The old sogger nearly played us a game of freeze-out," Comstock complained, before it finally rose, covered with mud, to take a breath.

When a bowhead stayed down this long, it was usually so exhausted and breathless that it could easily be lanced. But often a bowhead adopt-

"Load with blocks of granite to utmost content," read one order. Commanded another: "Have a pipe and valve fitted under skillful direction so that after anchoring in position the water can be readily let into the hold." No stranger articles of shipping were ever issued to Yankee whalers than those received by 39 ships purchased by the Union Navy in 1861, at the outset of the Civil War.

Yet the mission of the South Atlantic Blockading Squadron was plain enough: the whalers, which would become known as the Stone Fleet, were to be scuttled in the harbor channels of Savannah, Georgia, and Charleston, South Carolina, thus blockading two vital Confederate supply ports.

Herman Melville called it "the terrible Stone Fleet, on a mission as pitiless as the granite that freights it." Arriving off Savannah, the Union officers discovered that the Confederates had done their job for them. Mistaking the whalers with their painted gunports for an invasion force, the Rebels had hurriedly sunk three steamers in the port's harbor entrance.

The Stone Fleet next headed north to Charleston, where 25 of the whalers were scuttled to plug two main harbor channels. After the operation ended in January 1862, the New York *Herald* gloatingly reported that "Charleston, so far as any commerce is concerned, may be considered 'up country.'"

But Rebel blockade runners were soon finding their way around the sunken steamboats in Savannah's harbor channel, and within four months the Charleston harbor, too, was back in business. On May 11, 1862, the United States Coast Survey reported that the tides and currents had scoured a new channel 21 feet deep around the sunken wrecks there. As Melville lamented: "A failure, and complete, was your old Stone Fleet."

ed the more frustrating tactic of running for an ice floe, bellowing as it ran, and ducking under the ice. The whalemen had to cut the harpoon line to keep from smashing their boat and being tossed into water so cold no one could live in it more than a few minutes. It was partly because of this evasive bowhead tactic that whalers finally turned to the bomb lance, which exploded on penetration and hastened the kill. Even so, a great many more bowheads were seen than caught.

But they were worth the chase. Their blubber was twice as thick as that of the sperm whale; a large bowhead yielded as much as 300 barrels of oil. Though inferior to sperm oil, it was still in demand as a lubricant by cordage manufacturers, screw cutters and steel temperers, and as fuel for miners' lamps. The bowhead's 3,500 pounds of baleen was simply sliced from the jaw, scraped clean of its gummy covering—the whalemen called this process "knocking off the oysters"—and tied in bundles for shipment cross-country from San Francisco on the new railroad. Together the oil and baleen from a single bowhead could be worth $10,000.

The whalemen earned every penny of it. Even in the summer the Arctic was an eerie, frightening place. The seas were generally shallow and dotted with reefs that did not show up on the inaccurate maps. The magnetic influence of the North Pole induced wide compass deviations, which made navigation chancier still. On arctic duty the masthead lookout posts, swathed in canvas against the harsh elements, were for the first time called crow's-nests, and the sailors manning them were often fooled by mirages into calling out warnings of nonexistent land.

For two summer months the sun hardly set. Yet visibility was often zero because of banks of fog that rolled in and so blanketed the ships that men found themselves calling across the deck to shipmates they could not see. The foggy nights were sometimes so still that, as the whalemen said, "you could hear the moon shine, though you could not see it." Sometimes the night would erupt in noisy confusion when the soft swash of another whaler was heard approaching; horns were sounded, bells were rung, empty casks were pounded and guns were fired to warn the other ship and avoid collision.

A July day in the Bering Sea could have temperatures of 60° F., then freezing weather and heavy snow. A clear night sky could become a howling blizzard by dawn. But the worst menace, ever present, was ice.

From the time in June when the first whalers poked their bows into the channels of water among the ice floes to late September when ice films formed in the channels, the captains kept pushing farther north. That was where the bowheads went. They were sometimes called icebreakers by the whalemen; a bowhead could come up under ice two to three feet thick and bump its way through. Even the whaleships reinforced for arctic cruising, their bows double planked with tough white oak and sheathed with ⅞-inch cedar, could not punch through ice that thick.

The trick for the whalemen was to find the permanent ice pack as it retreated northward under the sun's warming rays. For here, just in front of the shelf, congregated the greatest numbers of bowheads. But it required a twisting, constantly tacking course through the broken ice floes spreading hundreds of miles before the shelf. And these floes, sometimes 10 to 20 feet thick, were themselves dangerous. Sometimes they

A New Bedford dock hand is dwarfed by sheaves of flexible baleen taken from the mouths of arctic bowhead whales. A bowhead yielded as many as 600 plates of the glossy baleen, measuring 1 to 14 feet from their blunt roots to their fringed ends, and worth up to five dollars per pound. New England manufacturers used them to make corset stays, umbrella ribs and buggy whips. The baleen from a single whale might bring as much as $15,000.

lay motionless on the calm waters; at other times they came down on the whalers, driven before a sudden storm, crushing everything in their path. The whaler that could not beat a hasty retreat was in dire trouble.

A second danger constantly on the whalemen's minds as the season advanced was the chance of a sudden cold spell freezing the narrow Bering Strait while they were north of it. If that happened, it would have the effect of snapping shut the door of a huge trap, catching and perhaps dooming an entire fleet. Then, of course, there were the awful storms that grew in malignant intensity as winter neared; even if a whaler escaped the encroaching ice, she could be destroyed by wind and wave, leaving her crew—those who survived—at the mercy of the Arctic. Yet because the bowheads always seemed to stay on, the whaling captains remained, risking more each year to linger until the last possible day.

In 1870 one whaler, the *Japan*, lost the gamble against the weather. Her New Bedford captain, Frederick Barker, finding both the whaling and the weather still good, stayed north of the Bering Strait until the 4th of October. On the 4th, after running south for a few days, he encountered a strong gale, with freezing temperatures and winds that nearly dismasted his ship even under shortened sail. For four days the *Japan* was battered by the storm, until Barker was no longer sure of his position. He thought only that he was somewhere off the Alaskan coast near the Bering Strait. In fact he was close to the Siberian side.

"The gale blew harder, if possible," Barker later reported, "attended by such blinding snow that we could not see half a ship's length. Near noon, the weather still continuing very thick, we discovered breakers on the lee bow, close aboard. The ship had been running under lower topsails and storm sails, but owing to the strength of the gale, was making racehorse speed. The helm was put starboard and braced up sharp, when we discovered breakers off the weather bow, close aboard. Just then, to add to our horror, a huge wave swept over the ship, taking off all the boats and sweeping the decks clean. Our situation was now most critical, death truly stared us in the face."

Another gigantic sea, greater than all the others, came roaring down on the *Japan*. It lifted her high and then slammed her down broadside in the rocky shallows, shattering timbers and toppling masts. But its very enormity proved the whalemen's salvation, for it had deposited the *Japan* beyond the worst force of the seas that followed. Miraculously, no one had been killed in the beaching, and the 30 men on board managed to clamber onto land. But their torment was just beginning.

A number of Chukchis, a non-Eskimo Siberian people, had gathered on the beach, and they did what they could to assist the castaways. Barker recounted how the Chukchis bundled him onto a sled and took him to their settlement. "Some distance from the beach, we passed on the way the bodies of sailors frozen to death. The air was piercing cold and several of my men, being unable to dry their clothes, had fallen by the wayside and died. I discovered that out of a crew of thirty, already eight had frozen to death. All were badly frostbitten."

Barker believed himself to be freezing to death by the time he and his rescuers reached the village huts. "I was carried in like a clod of earth, as I could not move hand or foot," he recalled. "The chief's wife, in whose hut I was, pulled off my boots and stockings and placed my frozen feet against her naked bosom to restore warmth and animation."

The Chukchis sheltered Barker and his surviving crewmen throughout the Siberian winter, while temperatures plunged to 50° below zero and food supplies of walrus meat and blubber ran dangerously low. In a letter, Barker wrote: "Should I ever come to the Arctic Ocean to cruise again, I will never catch another walrus, for these poor people along the coast have nothing else to live upon."

In New Bedford and San Francisco the *Japan*'s failure to return from the Arctic might have been taken as an omen that the whalemen were pushing their luck in the far north. But it was not. By May 1871 that year's entire arctic fleet—40 vessels—was at Siberia's Cape Thaddeus, on the Bering Sea, waiting at the edge of the ice for the pack to retreat north. The captains, like the *Japan*'s ill-fated Barker, were as determined as ever to hunt as far north as possible, and to remain there until driven south.

By June, the whalers had made scant progress north, and in fact had been pushed a bit south to Cape Navarin by the ice. But now the bowheads had begun to appear; six were caught, and in moderating weather at the end of June the whalers followed the bowhead herds through the Bering Strait into the Chukchi Sea. There they suffered their first casualty. The *Oriole* was holed by the shifting ice floes. Her captain took her into Plover Bay for repairs, but she was too badly stove in. She was broken up for her gear, and her crew was assigned to the other whalers.

Shortly after they had traversed the Bering Strait, the whalemen spotted the survivors of the *Japan*, led by Captain Barker. The castaways had run down to the beach and were waving frantically. With them were their saviors, the Chukchis, and these men of the north somberly warned that the fleet was in for a bad season. The ice pack would be slow in retreating north and would return south much earlier than normal.

The captains, who did not want to believe it, put the warning down to superstition. They would go north. And while they waited for the ice pack to retreat, they occupied themselves, despite Barker's pleas, by killing the walruses upon which the Chukchis depended for life. One vessel alone, the *Contest*, slew 400 of the creatures during June, rendering them into 300 barrels of oil.

But bowheads were the prime quarry and, to get at them, the fleet had to work farther north and across the Chukchi Sea to the Alaskan coast. On July 5 Captain Valentine Lewis of the *Thomas Dickason* met with Captain Benjamin Dexter of the *Emily Morgan* and his wife, Almira, and complained of the slow-moving ice and the nearly constant fogs. Later in the month they were passing the time by gamming again, when the ice did break enough for most of the fleet to follow a narrow strip of clear water along the east shore of Alaska's Cape Lisburne all the way up to lat. 69° 10''. Thirty-nine whalers crowded into the few areas of open sea.

For the rest of July the fleet sat and waited in the cold, with few whales in evidence. From the crow's-nests nothing could be seen but a vast shelf of ice stretching westward from Blossom Shoals.

On August 6 the ice pack moved farther north and the whalers followed it. Within a few days 20 whalers were north of Blossom Shoals as far as Wainwright Inlet, just below Point Barrow, high up in the Arctic Ocean and as far north as most of the whalers had ever been. Here too were many bowheads. In the narrow strip of water the whaleboats went to work, and soon the sky was sooty with the smoke of the tryworks.

The weather improved, and there was promise of a good season. More than 100 whaleboats zigzagged back and forth through a 20-mile strip of water between the coast and the ice pack offshore. Most of the whalers were in sight of one another, and the usually barren area now resembled a bustling harbor in New England, with captains' wives and families visiting back and forth, exchanging gifts, gossip and recipes. By August 10 the whalers had captured a number of bowheads and were filling their holds with oil and baleen. The next day they got their first scare.

A strong wind out of the west suddenly moved a large ice pack shoreward. Some whalers had to retreat so precipitously that they were forced to slip their cables and leave their valuable anchors under the advancing ice—a painful decision for a thrifty Yankee captain. Many of the far-flung whaleboats were cut off and had to be hauled across the ice floes. Captain William Kelley of the *Gay Head* noted that the ice formed solidly around his whaleboats in only half an hour, adding that "we were forced to drag twenty-six boats over it. Fourteen boats were collected on a single cake at one time."

For two days the ice pressed in on the whalers, some of which went aground as they retreated into shallow water. Then, as abruptly, the ice pack halted, leaving a clear strip of water as far north as Point Belcher. The area also contained plenty of whales, but the whaleboats could not get to them. On August 16 the 20 whalers were joined by 12 more, working their way north through a coastal channel that was still clear.

However, the next day the ice started to move in on them again. Captain Kelley, an experienced arctic whaleman, knew there was deeper water inside the shoals that extended from Wainwright Inlet to Point Belcher. If the fleet could get inside, the shoals might protect them from the encroaching ice. Sounding with his lead, he found a channel deep enough to cross the shoals. As he went through, he marked the channel by tossing cordwood logs over the side, anchored by bricks. The rest of the fleet followed him into the refuge, and the boats went out again.

Their hunting grounds were now more restricted than ever. Inside the shoals, up against the shore of Point Belcher, there was scarcely room to anchor, much less sail down to a whaleboat that had caught a bowhead. And the whales seemed to sense the safety of the ice edge outside the shoals. Nearly every time one was struck, it dashed for the ice and went under it. In five frustrating days the men of the *Emily Morgan* struck and lost two whales, found a herd of them, struck only one, which also escaped, and chased some more through a snowstorm without another strike. Then, on August 25, the whalers' luck changed.

The westerly winds that had been driving the ice onto them shifted into the east and piped up to gale force. Slowly the ice pack broke up and moved away from the shoals. The gale moderated and died, and for the next two days the weather was excellent. More whales were caught, and more whalers sent columns of smoke skyward from their tryworks.

During this time, with the whaleships close to the shore, Alaskan

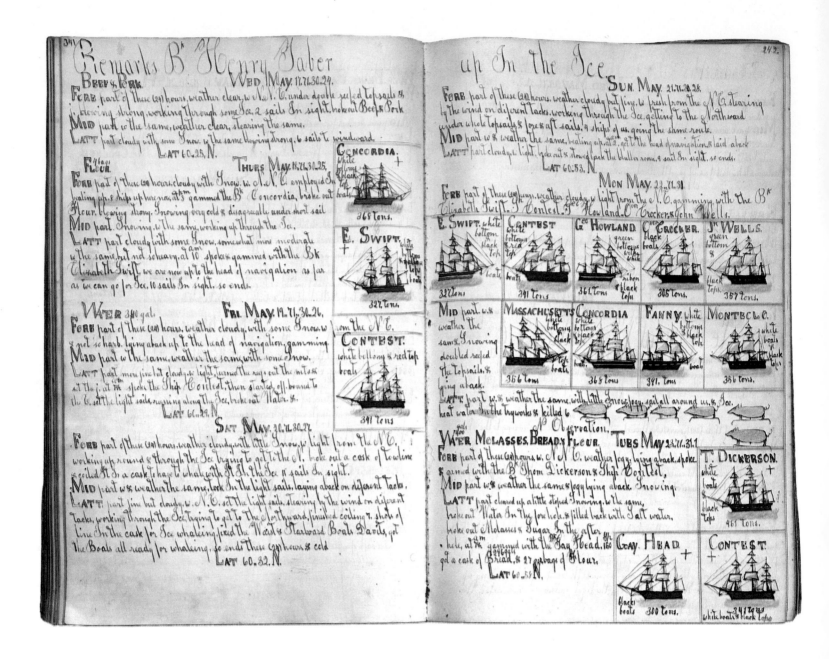

Eskimos came out to trade. Like the Chukchis, they advised the captains to take advantage of the open water to escape while there was still a chance. But the captains, some of them veterans of many summers in these waters, felt sure that they could count on another strong easterly to drive the ice to sea long enough for them to get out. Besides, there were hundreds of whales in the area. To leave in August, just on the Eskimos' say-so, they agreed, would be foolish. The whaling continued, and the whaling ships stayed where they were: 32 of them anchored in a 20-mile stretch of water barely half a mile wide and 20 feet deep.

On August 29 the wind shifted back into the west, pushing the ice pack toward the shoals off Point Belcher. Once again the deepest ice floes grounded on the shoals. Now there was more ice than ever, millions of tons of it, groaning and creaking; it slowly bumped over and

Two pages from a journal of Abram Briggs of the Henry Taber underscore the frustrations of arctic whaling. On May 18, 1871, wrote Briggs, the Yankee fleet reached "the head of navigation as far as we can go for Ice." He then related, with illustrations, how for six days the crews did very little but visit back and forth, and on one day roasted six pigs.

around the stranded floes and came down on the fleet resting at anchor.

The whalers let out more line and crowded up against the shore, some grounding close to the beach. The ice floes surrounded them, shoving them from side to side. By day the whalemen could watch the gradual filling up of their anchorage. At night they were kept awake by the grinding of the floes against the hulls. Every morning the skim ice that had formed in the night was thicker, filling the areas between the floes until the anchorage resembled the ice pack itself, beyond the shoals.

The first ship to fall victim to the ice was the *Comet* of Hawaii. Just before 3 a.m. on September 2, two huge floes converged on her, lifting her from the water and cracking her thick timbers as if they were walnuts. Her captain raised her ensign upside down, and at dawn, despite a heavy snowstorm, the nearby whalemen saw her distress signal. Whaleboats went over to her, their crews punching through the skim ice with their oars, and rescued captain and crew.

For three days the *Comet*'s wreckage perched crazily atop an ice floe; then the currents tipped the floe and sent the vessel's remains to the bottom. Still the captains optimistically awaited a northeaster to drive the ice back and free them. Working their way into one of the few open areas, Captain Lewis and a boat crew from the *Thomas Dickason* caught a whale. So did a crew from the *Roman*, which had tied up to a huge floe off the Seahorse Islands. The *Roman*'s men had begun to cut in their whales, when two more floes moved in on their ship.

The *Roman*'s captain, Jared Jernegan, saw the floes advancing on his vessel and watched in horror "as the ship was drove astern, the rudder fetching up against the ice. The ship's stern was all stove in by the heavy drift. Ice worked right under the ship, raised the whole ship almost out on the ice then her whole broad side was stove in."

Captain Jernegan ran down into his cabin, grabbed his chronometers and his pistol, and came back on deck as three whaleboats bumped over the side. He and his crew scrambled down into them. The ice under the *Roman*'s hull lifted her into the air, and the bark fell over onto her side and was crushed in the grip of the ice floes. Jernegan and his crew joined the other refugees being sheltered aboard the 30 remaining vessels.

The rest of the whalemen at last began to worry in earnest. Their concern increased on September 6, when the wind shifted again. This time it went into the southeast, which might have helped earlier, but at this point only compounded the trouble. There was too much ice inside the shoals for the wind to clear it out, and the gusts merely shoved the big cakes about, causing more damage. The *Eugenia* lost a section of copper on her bottom and some of her sheathing. The *John Wells* was driven aground; after an anxious night of pounding on the beach, her men had barely succeeded in kedging her off when they had to go to the aid of the *J. D. Thompson*, which had also gone aground. On September 8 the *Awashonks* became the third victim of the Arctic. She was crushed and carried aground by a gale. This one came from the south, driving the ponderous ice cakes against the other ships' hulls, breaking the *Eugenia*'s rudder and carrying the *Elizabeth Swift* aground.

The captains now gathered for a conference in the cabin of the *Gay Head*. All agreed that there was scarcely any hope that they could make it back across the shoals to the open sea. "Offshore is one vast expanse of ice," wrote the *Emily Morgan*'s mate William Earle. "Not a speck of water is to be seen in that direction." Still, the prospect of deserting their ships and their hard-won cargoes and risking the lives not only of the whaling crews but of wives and children was almost too much. "We felt keenly our responsibility," the *Gay Head*'s Captain Kelley later wrote, "with $3,000,000 worth of property and 1,200 lives at stake."

Nevertheless, there seemed no alternative. Every night as much as three inches of new ice formed in their cramped anchorage. Snowstorms were fiercer, and the landscape was becoming a silent white wilderness.

The whalemen dismissed the notion of wintering over. They had provisions for barely three months; the winter would last nine. And the Chukchis had been hard pressed to feed the crew of one whaler the year before; there was no chance that the Alaskan Eskimos could support the crews of 32 vessels. But how could they escape?

There were two possibilities. One was for the whalemen to work their way south inside the shoals in their whaleboats in an effort to find vessels free of the ice to rescue them. On September 10 Captain D. R. Fraser of the *Florida* set out with three boats to investigate this option.

Meanwhile, the remaining whalemen set to work on a second alternative—that they might somehow manage to sail south inside the shoals to freedom. The problem was not simply the encroaching ice floes; it was the shallowness of the inshore waters. The deepest channel south of Point Belcher was blocked by a bar that rose to within five feet of the surface, far too shallow for most of the heavily laden whalers. But perhaps the two smallest vessels in the fleet, the *Kohola* and the *Victoria*, could be lightened enough to clear the bar. If so, all the whalemen would crowd on board and they would attempt to sail south, with whaleboats leading the way and towing them out of tight spots in the floes.

The whalemen spent a tiring day unloading the two ships' cargoes and heavy gear. But neither could be lightened enough to clear the bar; all they could do was raise the *Kohola* to a nine-foot draft, still too deep.

Everything now depended on finding some vessels farther south that were free of the ice. Without rescue ships, it would be suicide to take their whaleboats into the stormy open waters of the Chukchi or the Bering Sea. In such an event, said Captain Dexter of the *Emily Morgan*, "out of the 1,200 men, not 100 will survive."

On September 12, however, Captain Fraser returned with heartening news. Some 80 miles away, off Icy Cape, just below the southern limit of the ice pack, were two whalers in open water. Five more were caught in the floes near the shelf but were working their way free.

The news that there were more than 1,000 of their comrades in a desperate situation had been a severe blow to the captains of these ships. Together the seven whalers had taken about $100,000 worth of oil and baleen; they were counting on much more as they worked their way down through the Bering Strait along with the whales. But their response to their fellow whalemen's pleas was immediate. They would stay where they were until the refugees could reach them.

Fraser had got to them through a narrow strip of water, clotted with ice floes but still passable by whaleboat. This last escape route was closing

The great arctic disaster

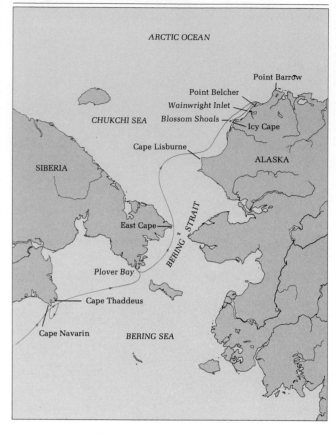

The worst peacetime whaling disaster of the 19th Century was the destruction of all but seven ships of a 40-vessel arctic whaling fleet in 1871. Gathering at Siberia's Cape Thaddeus about May 1, the bulk of the fleet pushed 1,000 miles north to Point Belcher, Alaska, where the whalemen pursued bowheads until September 14, when the ice closed in.

fast, with new ice forming every night. However, nearly everyone in the fleet was ready to go at a moment's notice. While awaiting Fraser's return, the men had been preparing the whaleboats by nailing "shoes"— planking reinforcement and copper sheathing—over the bows and keels, and by building up the bulwarks to provide extra freeboard. Some whaleboats were loaded with provisions, since the rescuers would obviously not have enough to feed 1,200 refugees. The rest would be jammed with about 14 people each, in space that was tight for the usual six.

Aboard every whaler the flag at the mizzen peak was lowered and raised again, union down as a sign of distress. In every forecastle the whalemen pulled on layers of pants and coats. In the aftercabins the captains took their chronometers from the bulkheads, packed a few precious possessions and damped their stoves. The captains' wives bundled together what little they could cram into the whaleboats. The children clutched their few indispensable dolls and toys. With a last painful look about their cabins, still intact and warm and full of souvenirs collected from all over the Pacific, the whaling families went on deck into the raw wind and climbed down into the bobbing, bumping whaleboats. Along the shore, whalemen could see groups of Eskimos waiting for the ice to be firm enough for them to go help themselves to the ships' stores.

Captain Thomas Williams, who had taken his wife, Eliza, to the Pacific aboard the *Florida* in 1858, was now in command of the *Monticello*, locked in the ice with the other whalers off Point Belcher. With him again were his wife; his son Willie, 12; and daughter, Mary, 10. Stancel, age 19, had also joined his family and was in the *Monticello*'s crew.

To the youthful Willie, the whole thing at first seemed like a grand adventure. When the *Monticello* was driven ashore by an ice floe and crews from the other whalers came on board to help kedge her free by hauling on anchors, Willie recalled in later life, "to me it was a gala day." His mood changed at departure. "It was depressing enough to me, and you know a boy can always see possibilities of something novel or interesting in most any change, but to my father and mother it must have been a sad parting, and I think what made it still more so was the fact that only a short distance from our bark lay the ship *Florida*, of which my father had been master eight years and on which three of his children had been born. The usual abandonment of a ship is the result of some irreparable injury and is executed in great haste; but here we were leaving a ship that was absolutely sound, that had been our home for nearly ten months and had taken us safely through many a trying time."

The flotilla of whaleboats set out early in the morning of the 14th and, by rowing and sailing all day in excellent weather, a number of them managed to reach Icy Cape just at dusk. Willie Williams recollected that "a tent was erected for the ladies and children and great fires were built for the men and for cooking."

The rescue vessels were in the open ocean, and there was real concern that some of the heavily loaded whaleboats might be swamped in the waves. The worry turned to fear next morning when the rest of the flotilla arrived in the midst of a heavy gale. "My father," Willie said, "had decided to go aboard the *Progress*. She was still at anchor and

pitching into the heavy seas, that were then running in a way that would have made you wonder how we would ever get the men aboard, let alone a woman and two children; but it was all accomplished without accident, or even the wetting of a foot.''

In other whaleboats the men bailed constantly as the seas poured over the gunwales. Everyone was soaked in freezing brine, and much of the bread, flour and other provisions so laboriously transported south were ruined by the salt water. But not one life was lost.

Besides the Williams family, the *Progress* took aboard two more captains' wives and two children, one of them a baby, and 188 whalemen. All seven whalers off Blossom Shoals had by now freed themselves, though the storm broke the *Arctic*'s port anchor cable and nearly drove her onto the ice before her starboard anchor caught and held. The *Arctic* rescued 250 people, the *Europa* 280, the *Lagoda* 195, the *Daniel Webster* 113, the *Midas* 100, the *Chance* 60. To many of the refugees the most poignant time of all came when they had to abandon the lifesaving whaleboats that had brought them through the ice and the gale winds. There was no room for them aboard the rescue ships, and their owners had to watch the empty boats go rocking and twirling downwind and crash against the ice pack to leeward.

But that was a small matter. Somehow nearly 1,200 whalemen and their families found corners to crowd into aboard the rescue vessels, and most of the survivors reached Honolulu by October 23, 1871.

Ironically, the Eskimos later reported that two weeks after the departure of the whalemen an intense northeaster swept across Point Belcher,

Six whaleboats rigged with sails bring some of the 1,200 men, women and children from their icebound whaleships to Alaska's Icy Cape in 1871. Many other whaleboats are drawn up on shore, and the refugees have put up tents and built huge bonfires for warmth and cooking.

Braving heavy seas in their overloaded boats, the crews of abandoned whalers make their way to seven rescue vessels anchored five miles off Alaska. One captain reported that a fierce "southwesterly gale tossed the whaleboats like pieces of cork." But not a single life was lost.

cleared the shoals of floes and drove the ice pack away from the abandoned fleet. But the pack soon returned, and then solid ice locked the whalers in for good. The patiently waiting Eskimos walked out to the ships and helped themselves to cordage, other gear and all the food left in the holds. When they found the medicine chests, some of them drank everything in the bottles, poisoning themselves and dying on the ice. The others, convinced that evil spirits were taking revenge, set fire to some of the ships.

When news of the loss of the ships reached New Bedford, the effect was cataclysmic. The Yankee whaling industry, crippled by the fires of the Civil War, now seemed all but destroyed by the ice of the Arctic. Most of the vessels were insured but their policies had been written by mutual insurance companies largely owned by New Bedford shipowners and whaling merchants who now had no money to pay out. And so the loss was total. Whalebone, which had once dropped to pennies per pound, soared to three dollars—because there was virtually none to be sold.

Whalemen as well as owners were ruined by the 1871 loss. From captains to cabin boys, they had all signed on for a share of the profit; since there was no profit, they received nothing for their summer's work. Nor were there berths on other whalers; there were now too many whalemen for the remaining ships. And the example of 1871 discouraged other bowhead hunters. Few made plans to return north in the spring.

But there was one man who was yearning to return to the Arctic. He was Captain Thomas Williams, late of the *Monticello*, and he had a daring plan to recoup some profit from the recent disaster. He formed a partnership in San Francisco with a number of daring investors led by Dr. Samuel Merritt, a man of many parts—among them physician, ship-owner, insurance underwriter and gambler. With Merritt's financial backing, Williams prepared to sail back to Point Belcher to salvage the oil and bone that had been left in the hulls of the ice-trapped whalers.

It was a gamble, since the ice might have broken up every ship and sunken every oil cask and slab of baleen. But there were 32 vessels up there, and Williams reasoned that a few of them, even if they were no longer seaworthy, might have escaped total destruction. If so, there was a treasure of oil and bone there, preserved in deep freeze, and it would command higher prices than ever before.

In the early spring of 1872, Williams bought the Hawaiian-registered bark *Florence*, outfitted her, signed on an experienced whaling captain, E. P. Herendeen, as first mate, and enlisted a crew. By mid-spring, earlier than whalers usually set out for the Arctic, Captain Williams—this time without Eliza and the younger children but with his older son, Stancel, on board—took the *Florence* out past the Golden Gate and set her course for the Bering Strait.

Before he got there he found he was engaged not only in a salvage expedition but also in a race. Three weeks before the *Florence* sailed, the bark *Francis Palmer*, under a Captain Jacobsen, had left San Francisco on the same mission. She followed by a week the fast schooner *Eustace*, commanded by Captain E. Everett Smith, an experienced arctic explorer. Both captains had given false destinations to the harbor master, and both were running under full sail for the Bering Strait.

Their early departure turned out to be in vain. When they reached the lower limits of the ice floes, just above the Bering Strait, they could only pick their way slowly through the narrow twisting leads of open water—and watch the *Florence* outsmart them. The *Eustace*, with its veteran arctic captain, was still narrowly ahead as the *Florence* approached the area where the fleet had been trapped. At that point, Williams lowered several whaleboats that swiftly maneuvered past the *Eustace*.

In short order the race was over. The whaleboats rowed on to find the fleet and claim possession, and soon afterward the *Florence* edged past the *Eustace* and arrived on the scene.

At first glance it seemed that Williams had come on a fool's errand. The area was a vast litter of broken hulls and twisted wreckage. Only a few of the abandoned vessels were recognizable. Of Williams' *Monticello* nothing was left but two sections—her stern, encased in a solid block of ice, and a piece of her iron-plated bow protruding from a floe half a mile away. The *Seneca* was a skeleton of ribs and bulkheads on the beach. The beached *Thomas Dickason* rested on her beam, full of water. The *Champion* was a jumble of wreckage on the shore. So were the *Emily Morgan*, the *Kohola* and the *Reindeer*. The *Awashonks* was found on the beach in two feet of water. The rest of the fleet was either burned beyond recognition or had gone to the bottom.

Only one vessel in the entire fleet, the *Minerva*, looked salvageable. Indeed, she was scarcely damaged. While the moving floes had smashed every other ship in the fleet, the water around the *Minerva* had evidently swiftly frozen into a giant floe, encasing and protecting her throughout the winter. There were signs that Eskimos had been aboard, but they had taken only some of her rigging and gear, and had not tried to burn her.

Further investigation disclosed more good news. The *Thomas Dickason*'s dismantled hull still protected her cargo. Many of the *Reindeer*'s casks, neatly stacked in her canted hull, were frozen in so solidly that they could not be broken out, but hundreds of other casks had gone ashore. Some were smashed, and the Eskimos had broken others open, but many remained sealed, waiting for Williams and his salvage crew.

While the *Eustace* and the *Francis Palmer* still battled to get through the moving ice south of the *Florence*, Williams set his men to work, filling the *Florence* and the remaining space in the *Minerva* with the casks and baleen from the wrecks and the shore. The Eskimos not only watched but helped, inviting Williams to their huts, which were luxuriously outfitted with the whalers' furniture and gear. Sail-canvas tents spread through the settlement. One shanty was paneled with fine wood from a ship's cabin. Williams had brought along some guns and trinkets, which he traded for baleen the Eskimos had taken from the wreckage.

Then, appointing his son Stancel captain of the *Minerva*, he led the way in the *Florence* down the coast that had been solid ice 10 months earlier. By now it was opening up for the few whalers poking their bows north for another season—and for the other two salvage ships that were too late. At last, in September of 1872, the *Florence* and the *Minerva* sailed triumphantly into San Francisco with 1,300 barrels of oil and $10,000 worth of bone, plus some walrus oil. There was some muttering that the captain should claim only a salvager's share of around 75 per cent. But when the question was put to the tough old whaleman, he roared: "Salvage? No! We intend to claim all that we got!" And so he did, for a grand total of nearly $20,000—considerably more than he had brought back from many a previous voyage.

Only a few stubborn whalemen went into the Arctic for the bowhead seasons in the years following 1871. Among them was Thomas Williams. And then in 1876 there was an almost exact recurrence of the events of 1871: the ice formed early, the captains waited for a northeaster to clear a path south, and an ice trap surrounded them. This time only 12 ships were caught. But some of the whalemen who tried to escape never made it. Nearly 50 men who attempted to winter on board their ships were lost. When the fleet returned next season, there was no trace of men or ships. One crew that abandoned ship and survived was that of Captain Williams. He and his crew escaped through the narrow channels in whaleboats as they had in 1871.

Williams tried only one more whaling cruise, and then went into the coal, lumber and carriage-making business in Oakland, California. He died there, at age 60, in 1880.

Another debacle occurred in 1888, when five more of the ships that went north were lost in the ice. By now, New England's whaling industry was entering its death throes. Petroleum oil had all but replaced whale oil as a primary fuel and lubricant, and in 1906 the invention of flexible spring steel replaced baleen for many uses.

Whaling itself did not end, of course. Markets for products still existed, provided costs could be kept low and volume high. What had once been a human adventure became in the 20th Century a sophisticated, highly mechanized industrial process, as first the Norwegians and then the Japanese and Russians, armed with modern weapons, waged relentless war on whales (page 33) for their oil—and increasingly in later years, their flesh. But Yankee whalemen played no part in this; by the time modern whaling developed, the New England industry was dead.

On the wharves of New Bedford the great ranks of oil casks lay rotting as lower petroleum prices dragged whale oil prices down. One crusty old merchant holding a supply of whale oil vowed that he would hold out until "sparm went back to a dollar." His cask hoops rusted, the staves shrank and the oil leaked away. A whaleship owner took his vessel up the Acushnet River and scuttled her, rather than watch her settle into the mud and lose her spars one by one in the dying forest of whaleship masts crowding the New Bedford wharves. Another whaleship was towed into the harbor and set afire in a bizarre Fourth of July holiday spectacle. But the watching crowd was silent, and the incident cast a pall over the city.

The old captains could sometimes be seen visiting the ghostly wharves, reminiscing with one another. On the rare instances when a whaler returned from a short Atlantic cruise, an occasional white-bearded whaleman might turn up to beg for some of her leftover salt beef, for old times' sake. Many retired captains retreated inland. Captain John Orrin Spicer of New London lived out his last years on a farm, a dotty old man stomping through his fields in Eskimo boots and a sombrero, and calling for warm seal's blood, which he believed to be good for his heart.

Not all of the old whaleships rotted at their wharves. The *Sunbeam* survived as a whaler until 1911, when she went into the packet trade, only to be wrecked among the Cape Verde Islands. The *Charles Hanson* was converted to power and put into the merchant service; carrying a cargo of dynamite, she caught fire off the Mexican coast in the early 1900s and ended her days in one glorious explosion.

Only a few stuck it out, sailing to their deaths still in pursuit of the whale. The *Swallow* was wrecked on Blyth Beach, Georgia, in January 1901. The *Alice Knowles* was sunk by a hurricane in the North Atlantic in 1917. The *Sea Ranger* was lost on Alaska's Kodiak Grounds in 1893. The *George and Susan,* named for shipowner George Howland and his wife, was wrecked in a gale off Alaska's Wainwright Inlet in 1885. The *Gay Head,* which bore the same name as one of the 1871 victims and was the last of the sailing whalers from San Francisco, was lost off Alaska in 1914. The last of the Yankee square-rigged whaleships to sail from New Bedford was the *Wanderer.* On August 25, 1924, she anchored in Buzzards Bay, Massachusetts, to await her crew. She was overwhelmed by a hurricane and piled onto Sow and Pigs rocks off Cuttyhunk Island.

Of them all, one alone survived, escaping the Confederate raiders, the ice of the Arctic and even the U-boats of World War I, to become the centerpiece of a maritime museum at Mystic, Connecticut. She was the *Charles W. Morgan,* from whose deck 18-year-old Nelson Haley, and hundreds of other Yankees for 80 years, embarked in little whaleboats to plunge into the "charmed, churned circle" of the giant sperm whale.

The Wanderer, last full-rigged whaler to sail from New Bedford, lies wrecked just 13 miles from her home port on August 26, 1924. She had departed the previous day, but had to anchor off Hen and Chickens Lightship while her captain finished last-minute business ashore. While she waited, a fierce northerly drove her onto the rocks just off Cuttyhunk Island, where the surf battered her to bits.

The Charles W. Morgan, the last survivor of the great 19th Century Yankee whalers under sail, lies at her berth in modern-day Mystic, Connecticut, a floating memorial to the great era of whaling. The Morgan, also portrayed on page 18 in full pursuit of her prey, hunted whales for 80 years, earning nearly two million dollars for her owners, before she was retired in 1921.

Bibliography

Allen, Everett S., *Children of the Light: The Rise and Fall of New Bedford Whaling and the Death of the Arctic Fleet.* Little, Brown, 1973.

Ashley, Clifford B., *The Yankee Whaler.* Houghton Mifflin, 1938.

Beale, Thomas, *The Natural History of the Sperm Whale.* John Van Voorst, 1839.

Cary, William S., *Wrecked on the Feejees.* Inquirer and Mirror Press, 1949.

Chatterton, E. Keble, *Whales and Whaling.* Gale Research, 1974.

Chippendale, Captain Harry Allen, *Sails and Whales.* Houghton Mifflin, 1951.

Craig, Adam Weir, *Whales and the Nantucket Whaling Museum.* Nantucket Historical Association, 1977.

Day, A. Grove, ed., *Mark Twain's Letters from Hawaii.* Appleton-Century, 1966.

Dozier, Thomas A., *Whales and Other Sea Mammals.* Time-Life Films, 1977.

Editors of American Heritage, *The Story of Yankee Whaling.* American Heritage, 1959.

Garner, Stanton, ed., *The Captain's Best Mate: The Journal of Mary Chipman Lawrence on the Whaler Addison, 1856-1860.* Brown University Press, 1966.

Haley, Nelson Cole, *Whale Hunt.* Ives Washburn, 1948.

Hare, Lloyd C. M., *Salted Tories; The Story of the Whaling Fleets of San Francisco.* Marine Historical Association, Inc., Mystic, Conn., 1960.

Haverstick, Iola, and Betty Shepard, eds., *The Wreck of the Whaleship Essex: A Narrative Account by Owen Chase.* Harcourt, Brace and World, 1965.

Henderson, David A., *Men and Whales at Scammon's Lagoon.* Dawson's Book Shop, 1972.

Hohman, Elmo P., *The American Whaleman.* Augustus M. Kelly, 1972.

Horan, James D., ed., *C.S.S. Shenandoah: The Memoirs of Lieutenant Commanding James I. Waddell.* Crown, 1960.

Kugler, Richard C., "The Penetration of the Pacific by American Whalemen in the 19th Century," *The Opening of the Pacific—Image and Reality.* National Maritime Museum, Greenwich, England, 1971.

Laing, Alexander, *Seafaring America.* American Heritage, 1974.

Leyda, Jay, *The Melville Log.* Gordian Press, 1969.

Liversidge, Douglas, *The Whale Killers.* Rand McNally, 1963.

Melville, Herman:
Moby Dick or, The Whale. Dell, 1972.
Omoo. Northwestern University Press, 1968.

Murphy, Robert Cushman:
A Dead Whale or a Stove Boat. Houghton Mifflin, 1967.
Logbook for Grace. Time Inc., 1965.

Olmsted, Francis Allyn, *Incidents of a Whaling Voyage.* Charles E. Tuttle, 1969.

Philp, J. E., *Whaling Ways of Hobart Town.* J. Walch & Sons, 1936.

Purrington, Philip F., *4 Years A-Whaling.* Barre Publishers, 1972.

Reynolds, Jeremiah N., *Mocha Dick or the White Whale of the Pacific.* Scribner's, 1932.

Sanderson, Ivar T., *Follow the Whale.* Little, Brown, 1956.

Sawtell, Clement Cleveland, *The Ship Ann Alexander of New Bedford.* Marine Historical Association, Inc., Mystic, Conn., 1962.

Scammon, Charles M., *The Marine Mammals of the North-western Coast of North America.* Dover, 1968.

Scheffer, Victor B., *The Year of the Whale.* Scribner's, 1969.

Smith, Gaddis, "Whaling History and the Courts," *The Log of Mystic Seaport.* Oct. 1978.

Stackpole, Edouard A.:
The Sea-Hunters. Bonanza Books, 1953.
Whales & Destiny. University of Massachusetts Press, 1972.

Starbuck, Alexander, *History of the American Whale Fishery from Its Earliest Inception to the Year 1876.* Argosy-Antiquarian, 1964.

Watson, Arthur C., *The Long Harpoon.* George H. Reynolds, 1929.

Whipple, A. B. C.:
Vintage Nantucket. Dodd, Mead, 1978.
Yankee Whalers in the South Seas. Charles E. Tuttle, 1973.

Whiting, Emma Mayhew, and Henry Beetle Hough, *Whaling Wives.* Houghton Mifflin, 1953.

Williams, Harold, ed., *One Whaling Family.* Riverside Press, 1964.

Acknowledgments

The index for this book was prepared by Gisela S. Knight. The editors wish to thank Roy H. Andersen, artist *(pages 34-41)*, John Batchelor, artist *(pages 91-94)*, William A. Baker, consultant *(pages 34-41, 91-94)*, William L. Hezlep, artist *(page 164)*, Peter McGinn, artist *(end-paper maps)*, Jay H. Matternes, artist *(pages 29-31)*, and Thomas J. McIntyre, consultant *(pages 29-31)*.

The editors also wish to thank: In Germany: Arnold Kludas, Deutsches Schiffahrtsmuseum, Bremerhaven; Jürgen Meyer, Altonaer Museum, Hamburg. In Italy: Luisa Secchi, Director, Museo Navale, Genoa-Pegli; Commander Flavio Serafini, Ministero della Marina, Rome; Baron Giambattista Rubin de Cervin, Director, Museo Navale, Venice. In Japan: Yosuke Nakajima, Editor, Heibonsya Ltd.; Atsushi Fujiwara, Taiyo Fishery Company, Ltd.; Hideo Omura, Director, Whales Research Institute, Tokyo. In the Netherlands: Thijs Mol, Arnhe. In Paris: Denise Chaussegroux, Researcher; Hervé Cras, Director for Documentary Studies, Musée de la Marine. Elsewhere in France: Jacques Soulaire, Bourg-la-Reine; Jacques Kuhnmunch, Curator, Musée des Beaux-Arts, Dunkirk. In Wales: A. D. Fraser Jenkins, Assistant Keeper, R. G. Keen, Assistant Keeper, National Museum of Wales, Cardiff.

The editors also wish to thank: In Washington, D.C.: Ormond Seavey, Assistant Professor of English, The George Washington University; Michael Music, Navy and Old Army Branch, National Archives; Thomas R. Loughlin, Marine Mammal Research Specialist, National Marine Fisheries Service; Michael R. Harris, Museum Specialist, Division of Medical Sciences, The National Museum of History and Technology, Smithsonian Institution. Elsewhere in the United States: Patrick Dempsey, Gary Fitzpatrick, Geography and Map Division, Library of Congress, Alexandria, Virginia; Ralph Carpentier, Director, Town Marine Museum, East Hampton Historical Society, Amagansett, Long Island; Murray Morgan, Auburn, Washington; Lawrence Dinnean, Curator of Pictorial Collections, William M. Roberts, Assistant Head of Public Services, The Bancroft Library, University of California, Berkeley; Muriel C. Crossman, Librarian, Thomas Norton, Director, Dukes County Historical Society, Edgartown, Massachusetts; Debra Sullivan, Assistant to the Photo Librarian, Bishop Museum, Jean Martin, Lecturer, University of Hawaii Community Colleges, Honolulu; Henry Beetle Hough, Martha's Vineyard, Massachusetts; Willits D. Ansel, Georgia Hamilton, Head Cataloguer, Mary R. Maynard, Office of Public Affairs, Mystic Seaport, Inc., Mystic, Connecticut; Louise Hussey, Librarian, The Foulger Museum, Adam Weir Craig, International Marine Archives, Inc., Leroy H. True, Nantucket Historical Association, Nantucket, Massachusetts; Ronald Hansen, The Mariner's Home, Marge Habicht, Assistant Librarian, *The Standard-Times*, Philip Purrington, Nick Whitman, New Bedford Whaling Museum, New Bedford, Massachusetts; Edgar Mayhew, Lyman Allyn Museum, New London, Connecticut; Susan Hartzband, Research Assistant, Elizabeth Meisner, Registrar, Carolyn Ritger, Photographs Librarian, The Mariners Museum of Newport News, Virginia; Miwa Kai, Head, Japanese Section, East Asian Library, Columbia University, New York, New York; David A. Henderson,

Professor of Geography, California State University, Northridge, California; Francis F. Jones, Palo Alto, California; Barbara Johnson, Princeton, New Jersey; Virginia M. Adams, Curator, Special Collections, Providence Public Library, Providence, Rhode Island; Kathy Flynn, Photographic Assistant, Mark Sexton, Staff Photographer, Philip C. F. Smith, Curator of Maritime History, Peabody Museum, Salem, Massachusetts; Kenneth R. Martin, Director, Kendall Whaling Museum, Sharon,

Massachusetts.

Quotations from *One Whaling Family*, edited by Harold Williams, Riverside Press, © 1964 Houghton Mifflin Company; from *The Captain's Best Mate: The Journal of Mary Chipman Lawrence on the Whaler Addison, 1856-1860*, edited by Stanton Garner, © 1966 Brown University Press; from *Whale Hunt*, by Nelson Cole Haley, © 1948 by Ives Washburn (now David McKay); and from *The Ship Ann Alexander of New Bedford* by Clement Cleveland Saw-

tell, published by The Marine Historical Association, Inc. (now Mystic Seaport Museum, Inc.), 1962, are reprinted by permission from their publishers. Quotations from *Whaling Wives*, by Emma Mayhew Whiting and Henry Beetle Hough, published by Houghton Mifflin Company, 1953, reprinted by permission of Mr. Hough. A particularly valuable source of quotations was *The Marine Mammals of the North-western Coast of North America* by Charles M. Scammon, Dover, 1968.

Picture Credits

The sources for the illustrations in this book are shown below. Credits from left to right are separated by semicolons, from top to bottom by dashes.

Cover: New Bedford Whaling Museum. Front and back end papers: Drawing by Peter McGinn.

Page 3: Paulus Leeser, courtesy Barbara Johnson Collection. 6, 7: Library of Congress. 8-15: The Kendall Whaling Museum, Sharon, Mass. 16: Peabody Museum of Salem. 18: *The Charles W. Morgan*, by John F. Leavitt, courtesy Mystic Seaport Museum, Inc., Mystic, Conn. 19: Courtesy Mystic Seaport Museum, Inc., Mystic, Conn. 20: New Bedford Whaling Museum. 21: Peabody Museum of Salem. 22, 23: New Bedford Whaling Museum. 25: Courtesy The Mariners Museum of Newport News, Va. 27: New Bedford Whaling Museum. 29-31: Drawings by Jay H. Matternes. 34-41: Drawings by Roy Andersen. 42, 43: New Bedford Whaling Museum. 44: Scala, courtesy Cappella degli Scrovegni, Padua. 45: Courtesy Private Collection. 48: New York State Historical Association, Cooperstown. 49: The Kendall Whaling Museum, Sharon, Mass. 51: New Bedford Whaling Museum. 54: Henry Beville, courtesy Nantucket Historical Association. 55: Courtesy American Philosophical Society. 56: Courtesy National Museum of Wales, Cardiff. 58: Henry Beville, courtesy Nantucket Historical

Association. 61-65: New Bedford Whaling Museum. 66, 67: Henry Beville, courtesy Nantucket Whaling Museum. 68, 69: New Bedford Whaling Museum. 70, 71: Library of Congress. 72, 73: Henry Beville, courtesy Nantucket Historical Association. 74: Seamen's Bethel, sponsored by New Bedford Port Society—New Bedford Whaling Museum. 77: New Bedford Whaling Museum. 78: Henry Beville, courtesy Nantucket Whaling Museum. 79: Library of Congress. 81: New Bedford Whaling Museum. 82: Paulus Leeser, courtesy Barbara Johnson Collection. 84: New Bedford Whaling Museum. 86: Forbes Collection, Hart Nautical Museum, M.I.T. 88: By permission of the Houghton Library, Harvard University. 91-94: Drawings by John Batchelor. 96, 97: Bernice P. Bishop Museum, Honolulu. 98: Peabody Museum of Salem. 100-107: New Bedford Whaling Museum. 108, 109: Library of Congress. 110: New Bedford Whaling Museum. 112, 113: Courtesy Dukes County Historical Society, Edgartown, Mass. 116-118: Paulus Leeser, courtesy Barbara Johnson Collection. 119: Nicholson Whaling Collection, Providence Public Library. 121: Library of Congress. 122, 123: New Bedford Whaling Museum. 124: Barbara Johnson Collection—courtesy Stanton Garner. 125: Palmer Fund, Lyman Allyn Museum, New London, Conn. 126: Courtesy Mystic Seaport Museum, Inc., Mystic, Conn. 127: New

Bedford Whaling Museum. 128: Paulus Leeser, courtesy Barbara Johnson Collection (2); New Bedford Whaling Museum—Paulus Leeser, courtesy Barbara Johnson Collection—New Bedford Whaling Museum. 129: Paulus Leeser, courtesy Barbara Johnson Collection; New Bedford Whaling Museum (2); Paulus Leeser, courtesy Barbara Johnson Collection (3)—courtesy The Mariners Museum of Newport News, Va. (2); Paulus Leeser, courtesy Barbara Johnson Collection (2). 130: Paulus Leeser, courtesy Barbara Johnson Collection. 131: Paulus Leeser, courtesy Barbara Johnson Collection, except top left, New Bedford Whaling Museum. 132, 133: Courtesy The Bancroft Library. 135, 137: Library of Congress. 138, 141: Courtesy The Bancroft Library. 143: Peabody Museum of Salem. 145: Library of Congress. 148, 149: Courtesy Kennedy Galleries, New York. 151: Chicago Historical Society. 153: New Bedford Whaling Museum. 154: The Mansell Collection, London. 156: Library of Congress. 159: *The Yankee Whalers*, by Clifford W. Ashley, © renewed 1955 by Sarah Ashley Delano, reprinted by permission of Houghton Mifflin Company. 162: Paulus Leeser, courtesy Barbara Johnson Collection. 164: Map by William Hezlep. 166-170: New Bedford Whaling Museum. 171: Oliver Denison III, courtesy Mystic Seaport Museum, Inc., Mystic, Conn.

Index

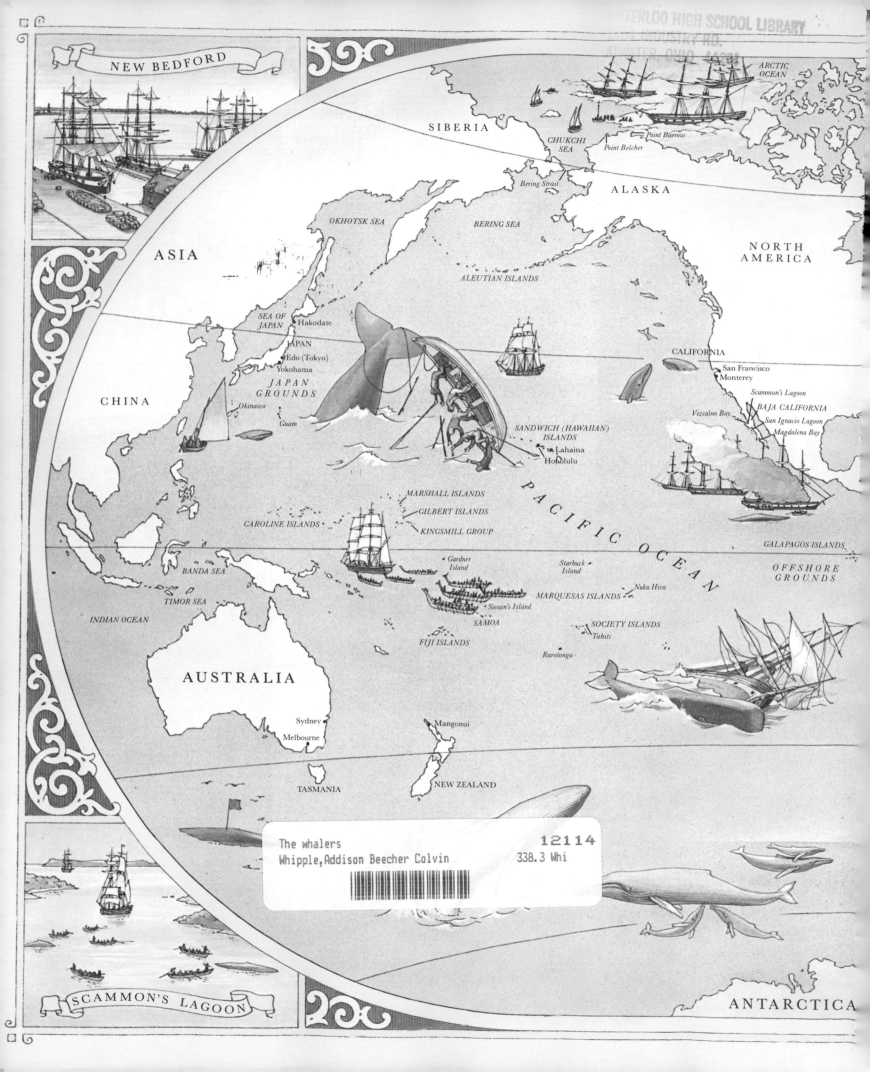

NEW BEDFORD

SIBERIA

ARCTIC OCEAN

CHUKCHI SEA

Point Barrow

Point Belcher

Bering Strait

ALASKA

OKHOTSK SEA

BERING SEA

NORTH AMERICA

ASIA

ALEUTIAN ISLANDS

SEA OF JAPAN

Hakodate

JAPAN

Edo (Tokyo)

Yokohama

CALIFORNIA

San Francisco

Monterey

JAPAN GROUNDS

Scammon's Lagoon

CHINA

Okinawa

BAJA CALIFORNIA

Vizcaino Bay

San Ignacio Lagoon

Guam

Magdalena Bay

SANDWICH (HAWAIIAN) ISLANDS

Lahaina

Honolulu

PACIFIC OCEAN

MARSHALL ISLANDS

GILBERT ISLANDS

CAROLINE ISLANDS

KINGSMILL GROUP

GALAPAGOS ISLANDS

Gardner Island

Starbuck Island

OFFSHORE GROUNDS

BANDA SEA

Nuku Hiva

MARQUESAS ISLANDS

TIMOR SEA

Swain's Island

INDIAN OCEAN

SAMOA

SOCIETY ISLANDS

Tahiti

FIJI ISLANDS

Rarotonga

AUSTRALIA

Sydney

Melbourne

Mangonui

TASMANIA

NEW ZEALAND

SCAMMON'S LAGOON

ANTARCTICA